Biomass and Biofuels

Biomass and Biofuels

Mason Jones

www.callistoreference.com

Callisto Reference,
118-35 Queens Blvd., Suite 400,
Forest Hills, NY 11375, USA

Visit us on the World Wide Web at:
www.callistoreference.com

ISBN: 978-1-64116-224-1 (Hardback)

Cataloging-in-Publication Data

Biomass and biofuels / Mason Jones.
    p. cm.
Includes bibliographical references and index.
ISBN 978-1-64116-224-1
1. Biomass. 2. Biomass energy. 3. Biology. 4. Energy conversion. I. Jones, Mason.
TP339 .B56 2019
662.88--dc23

Table of Contents

Preface

Biomass refers to plant and plant-based materials that do not have application as food or feed, but are used for the generation of energy through combustion or via conversion into different forms of biofuel. It is considered a renewable source of energy. The conversion of biomass into biofuel is achieved by various chemical, biochemical and thermal methods. Some of the chemical constituents of biomass are cellulose, hemicellulose and lignins. Biofuels are derived from domestic, agricultural, commercial and industrial wastes. These can be solid, liquid or gaseous. Some examples are bioethanol, biodiesel, syngas, ethanol fuels, etc. Some of the diverse topics covered in this book address the varied aspects of biomass and biofuels. It elucidates new production techniques and their applications in a multidisciplinary approach. For someone with an interest and eye for detail, this book covers the most significant topics in this field.

A detailed account of the significant topics covered in this book is provided below:

Chapter 1- A biofuel is a fuel derived from plants or from agricultural, domestic or industrial wastes. This chapter has been carefully written to provide an easy understanding of biofuel and its generations. The topics included in this chapter on first generation, second generation, third generation and fourth generation biofuels are crucial for a complete understanding of the subject.

Chapter 2- Biomass refers to the generation of energy by burning wood and other organic matter. Biomass is directly used via combustion to generate heat or converted into various forms of biofuel. The topics elaborated in this chapter on woody and non-woody biomass will help in providing a better perspective of biomass and its various types.

Chapter 3- Various fuels can be produced with the aid of first, second, third or fourth generation bio-fuel production methods. Some important examples of biofuels are biogas, syngas, biodiesel, green diesel, bioalcohol, bioether, solid biofuel, etc. which have been extensively discussed in this chapter.

Chapter 4- Science and technology has undergone tremendous development in the past decade, which has resulted in the production of various biofuels. Some of the common sources of biofuels are corn, sugarcane, cellulose, Camelina, Jatropha, rapeseed, algal oil, etc. which have been extensively examined in this chapter.

Chapter 5- In order to completely understand biofuel chemistry and production, it is vital to understand the processes related to it. The following chapter elucidates the various methods of production of biodiesel, bioethanol, biogas and aviation biofuel as well as includes a detailed discussion of ultra- and high-shear in-line and batch reactors, supercritical process and ultrasonic reactor method.

It gives me an immense pleasure to thank our entire team for their efforts. Finally in the end, I would like to thank my family and colleagues who have been a great source of inspiration and support.

Mason Jones

Chapter 1

Biofuels and Generations of Biofuels

A biofuel is a fuel derived from plants or from agricultural, domestic or industrial wastes. This chapter has been carefully written to provide an easy understanding of biofuel and its generations. The topics included in this chapter on first generation, second generation, third generation and fourth generation biofuels are crucial for a complete understanding of the subject.

The standard definition of biofuel describes it as a renewable source of energy produced from organic matter. It is a fuel produced from biological resources that we have access to on a large scale, which makes it a very promising substitute for fossil fuels.

Current technology makes it possible for us to use a wide range of natural materials, such as wood-based products, agricultural and municipal waste and a multitude of crops, to obtain energy. You can power anything from your car to your computer with green energy developed from the environment.

The total amount of plants and natural materials that can be used to create energy is referred to as biomass. As for the plants that are grown with the specific aim of creating biofuel, they are called lignocellulosic materials.

Components of Biofuel

Biofuel is developed from biomass by employing a series of highly effective technologies. Let us tell you about the main types of biological matter that can be used to create power.

Plants

Plants are the core material that can be used to obtain biofuel. They can be used in very many ways to develop different types of fuels. Take a look at the main categories of plants that we can use to produce green energy.

Agriculture Residue

The agriculture industry generates an important amount of residue. But instead of spending money to destroy it, modern day farmers can actually use this type of waste to create green energy to power part of their farm. And it is an excellent way of putting waste to very good use.

The principle is quite simple. By burning the plants, we can create heat, as any person who enjoys camping knows. Then we can use this heat to boil water in a tank. This is important because this way we can create steam, which we can then use to spin a turbine.

Our turbine will thus be able to power a generator that can ultimately create electrical energy. This can then be used to power any electric device you want. The only difference is that producing it is far less harmful to the environment than burning fossil fuels and that we can expect to have access to plants for years and years to come.

Energy Crops

Energy crops are grown with the precise aim of being used as power sources. Imagine entire fields of elephant grass or soybean that can be used to create electricity. These are called lignocellulosic materials and they have been carefully selected so that they can generate the best kind of energy with the smallest investment.

Here are the most important criteria that an efficient energy crop needs to meet:

- It must have a rapid growth rate.
- It must have a high photosynthesis rate, in order to be able to generate as much heat as possible.
- It must be resistant to pests and parasites.
- It must be resistant to a wide range of environmental factors.

Wood and Forestry Remains

Wood is one of the most efficient biofuel sources. Certain tree species, such as willows or poplars, can actually be used as energy crops because they meet the most important eligibility criteria mentioned above. However, woody materials are mainly used as energy sources by the forestry industry.

Imagine the amount of waste and remains generated by this industry. Tree bark, wood chippings or sawdust can be used to create energy, which is actually cost efficient for forestry plants. They can create part of the power they need for their equipment by using their waste.

Waste

Waste, garbage, trash, rubbish, call it what you will and you will think of one thing only: useless. Each of us throws away an impressive amount of things every single day because we do not need it anymore. This leads to gargantuan quantities of waste that can be harmful to our environment.

But what if we could take all of this rubbish and put it to good use? Well, creating energy from Municipal Solid Waste (MSW) is one of the best uses available for it today. Burning the MSW generates heat, which can then be used to obtain energy.

The main shortcoming of this method is the fact that it generates a noteworthy amount of emissions that are harmful to the environment. This happens because the MSW includes not only plant-based material but plastic and other petroleum-based items.

This is why recycling is so important to our society. By separating our trash, each of us can contribute to a greener tomorrow because every type of waste can be reused properly. A few minutes can actually make a difference in this situation.

Biogas

Biogas can be obtained from both animal and plant-based materials. It is generated by the large amount of microorganisms as they feed on decaying plants and waste. This is a natural process that happens all around us. And we can actually use it to our advantage. Here are some of the most important biogas-generators:

Animal Waste

Farms produce sizable amounts of animal waste, which can be used to create energy. Current laws actually dictate that this waste be placed into digesters. These are special containers that stimulate the development of the microorganisms, thus creating more gas. The biogas can be collected through a pipeline from the digesters and then be used to create energy.

Landfills

Landfills are an obsolete method of disposing of our waste. We currently have much better ways of storing it than burying it underground. However, another thing that we do have is a considerable amount of landfills that we need to get rid of.

Collecting the gas from these landfills is the perfect start for this process. By inserting a pipeline inside the landfill, all the gas produced in this confined space can be transferred to processing units where it can be filtered so that it can eventually be used to generate energy.

Oil

Both vegetable and animal oil can be used as a prime source of biodiesel. This is extremely important because this type of biofuel is highly compatible with a wide range of internal combustion engines that are currently in use.

There are two main types of biofuels – ethanol and biodiesel. The simplest way to distinguish between the two is to remember ethanol is an alcohol and biodiesel is an oil. Ethanol is an alcohol formed by fermentation and can be used as a replacement for, or additive to, gasoline whereas biodiesel is produced by extracting naturally occurring oils from plants and seeds in a process called transesterification. Biodiesel can be combusted in diesel engines.

Biofuels are grouped by categories - first generation, second generation, and third generation – based on the type of feedstock (the input material) used to produce them.

- First generation biofuels are produced from food crops. For ethanol, feedstocks include sugar cane, corn, maize, etc. For biodiesel, feedstocks are naturally occurring vegetable oils such as soybean and canola.

- Second generation biofuels are produced from cellulosic material such as wood, grasses, and inedible parts of plants. This material is more difficult to break down through fermentation and therefore requires pre-treatment before it can be processe.

- Third generation biofuels are produced using the lipid production from algae.

In addition, the term "Advanced Biofuels" is used to describe the relatively new technological field of biofuel production that uses waste such as garbage, animal fats, and spent cooking oil to produce liquid fuels.

Biofuels are not as energy dense as conventional transportation fuels. 1 gallon of biodiesel has 93% of the energy of 1 gallon of diesel and 1 gallon of ethanol (E85) has 73% of the energy of 1 gallon of gasoline.

Economic and Environmental Considerations

In evaluating the economic benefits of biofuels, the energy required to produce them has to be taken into account. For example, the process of growing corn to produce ethanol consumes fossil fuels in farming equipment, in fertilizer manufacturing, in corn transportation, and in ethanol distillation. In this respect, ethanol made from corn represents a relatively small energy gain; the energy gain from sugarcane is greater and that from cellulosic ethanol or algae biodiesel could be even greater.

Biofuels also supply environmental benefits but, depending on how they are manufactured, can also have serious environmental drawbacks. As a renewable energy source, plant-based biofuels in principle make little net contribution to global warming and climate change; the carbon dioxide (a major greenhouse gas) that enters the air during combustion will have been removed from the air earlier as growing plants engage in photosynthesis. Such a material is said to be "carbon neutral." In practice, however, the industrial production of agricultural biofuels can result in additional emissions of greenhouse gases that may offset the benefits of using a renewable fuel. These emissions include carbon dioxide from the burning of fossil fuels during the production process and nitrous oxide from soil that has been treated with nitrogen fertilizer. In this regard, cellulosic biomass is considered to be more beneficial.

Land use is also a major factor in evaluating the benefits of biofuels. The use of regular feedstock, such as corn and soybeans, as a primary component of first-generation biofuels sparked the "food versus fuel" debate. In diverting arable land and feedstock from the human food chain, biofuel production can affect the economics of food price and availability. In addition, energy crops grown for biofuel can compete for the world's natural habitats. For example, emphasis on ethanol derived from corn is shifting grasslands and brushlands to corn monocultures, and emphasis on biodiesel is bringing down ancient tropical forests to make way for oil palm plantations. Loss of natural habitat can change the hydrology, increase erosion, and generally reduce biodiversity of wildlife areas. The clearing of land can also result in the sudden release of a large amount of carbon dioxide as the plant matter that it contains is burned or allowed to decay.

Some of the disadvantages of biofuels apply mainly to low-diversity biofuel sources—corn, soybeans, sugarcane, oil palms—which are traditional agricultural crops. One alternative involves the use of highly diverse mixtures of species, with the North American tallgrass prairie as a specific example. Converting degraded agricultural land that is out of production to such high-diversity biofuel sources could increase wildlife area, reduce erosion, cleanse waterborne pollutants, store carbon dioxide from the air as carbon compounds in the soil, and ultimately restore fertility to degraded lands. Such biofuels could be burned directly to generate electricity or converted to liquid fuels as technologies develop.

The proper way to grow biofuels to serve all needs simultaneously will continue to be a matter of much experimentation and debate, but the fast growth in biofuel production will likely continue. In the United States the Energy Independence and Security Act of 2007 mandated the use of 136 billion litres (36 billion gallons) of biofuels annually by 2022, more than a sixfold increase over 2006 production levels. The legislation also requires, with certain stipulations, that 79 billion litres (21 billion gallons) of the total amount be biofuels other than corn-derived ethanol, and it continued certain government subsidies and tax incentives for biofuel production.

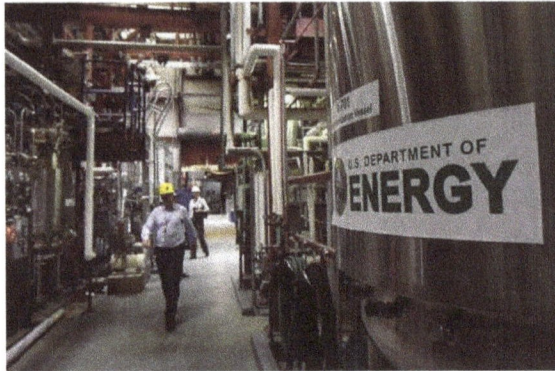
biofuels testing centre

One distinctive promise of biofuels is that, in combination with an emerging technology called carbon capture and storage, the process of producing and using biofuels may be capable of perpetually removing carbon dioxide from the atmosphere. Under this vision, biofuel crops would remove carbon dioxide from the air as they grow, and energy facilities would capture the carbon dioxide given off as biofuels are burned to generate power. Captured carbon dioxide could be sequestered (stored) in long-term repositories such as geologic formations beneath the land, in sediments of the deep ocean, or conceivably as solids such as carbonates

Uses of Biofuel

Biofuel can be used a prime source of energy and heat. Thus, it can be easily integrated into our day to day lives because it is highly accessible. We have listed the main uses of biofuel so that you can visualize the flexibility of this natural resource.

Heat

One of the main ways of generating energy from biomass is combustion. Therefore, heat is actually obtained first in this process. By inserting a hot water pipeline into a building, biomass can be used successfully as a heat source.

Electrical Devices

Biomass can be used to create electrical energy, which can then be used to power any electrical device. The quality of the energy is impeccable and the entire production process is far less harmful to the environment.

On the other hand, the costs of obtaining energy from biomass are still quite high. But as the importance of biomass energy becomes more and more evident to society, the production process will be perfected and made much more accessible and efficient.

Internal Combustion Engines

Alcoholic biofuels and oil-based biodiesels can be used to power internal combustion engines. These are the main alternatives to petroleum-based fuels. For the time being, these biofuels can be mixed with standard gasoline and utilized for a wide variety of the car engines.

First Generation Biofuels

First Generation biofuels are produced directly from food crops by abstracting the oils for use in biodiesel or producing bioethanol through fermentation. Crops such as wheat and sugar are the most widely used feedstock for bioethanol while oil seed rape has proved a very effective crop for use in biodiesel. However, first generation biofuels have a number of associated problems. There is much debate over their actually benefit in reducing green house gas and co2 emissions due to the fact that some biofuels can produce negative Net energy gains, releasing more carbon in their production than their feedstock's capture in their growth. However, the most contentious issue with first generation biofuels is 'fuel vs food'. As the majority of biofuels are produced directly from food crops the rise in demand for biofuels has lead to an increase in the volumes of crops being diverted away from the global food market. This has been blamed for the global increase in food prices over the last couple of years.

First generation biofuels account for a meaningful share of transport fuel use. By volume, ethanol represents approximately 10% of the U.S. gasoline pool, and biodiesel about 3% of the diesel pool. Total production in 2014 was 14.3 billion gallons of ethanol and 1.4 billion gallons of biodiesel.[4,5] The market for first generation biofuels has been relatively flat the past few years due to practical limitations on how much biofuel can be blended with conventional fuel without requiring vehicle modifications. Ethanol can be used in existing gasoline internal combustion engines in blends up to 10% (E10), although extensive testing on blends up to 15% (E15) has shown that cars built in 2001 or later can use this higher blend. So-called flex-fuel vehicles (FFVs) can burn a mixture as high as 85% ethanol (E85), and over 17 million FFVs are on the road today. Biodiesel is typically used in blends with conventional diesel in mixtures up to 20% (B20), for both vehicles and in home heating oil (called "bioheat").

Some of the most popular types of first generation biofuels are:

Biodiesel: This is the most common type of biofuel commonly used in the European countries. This type of biofuel is mainly produced using a process called transesterification. This fuel if very similar to the mineral diesel and is chemically known as fatty acid methyl. This oil is produced after mixing the biomass with methanol and sodium hydroxide. The chemical reaction thereof produces biodiesel. Biodiesel is very commonly used for the various diesel engines after mixing up with mineral diesel. Now in many countries the manufacturers of the diesel engine ensure that the engine works well even with the biodiesel.

Vegetable oil: These kinds of oil can be either used for cooking purpose or even as fuel. The main fact that determines the usage of this oil is the quality. The oil with good quality is generally used for cooking purpose. Vegetable oil can even be used in most of the old diesel engines, but only in warm atmosphere. In most of the countries, vegetable oil is mainly used for the production of biodiesel.

Biogas: Biogas is mainly produced after the anaerobic digestion of the organic materials. Biogas can also be produced with the biodegradation of waste materials which are fed into anaerobic digesters which yields biogas. The residue or the by product can be easily used as manure or fertilizers for agricultural use. The biogas produced is very rich in methane which can be easily recovered through the use of mechanical biological treatment systems. A less clean form of biogas is the landfill gas which is produced by the use of naturally occurring anaerobic digesters, but the main threat is that these gases can be a severe threat if escapes into the atmosphere.

Bioalcohols: These are alcohols produced by the use if enzymes and micro organisms through the process of fermentation of starches and sugar. Ethanol is the most common type of bioalcohol whereas butanol and propanol are some of the lesser known ones. Biobutanol is sometimes also referred to as a direct replacement of gasoline because it can be directly used in the various gasoline engines. Butanol is produced using the process of ABE fermentation, and some of the experiments have also proved that butanol is a more energy efficient fuel and can be directly used in the various gasoline engines.

Syngas: This is a gas that is produce after the combined process of gasification, combustion and pyrolysis. Biofuel used in this process is converted into carbon monoxide and then into energy by pyrolysis. During the process, very little oxygen is supplied to keep combustion under control. In the last step known as gasification the organic materials are converted into gases like carbon monoxide and hydrogen. The resulting gas Syngas can be used for various purposes.

Second Generation Biofuels

Second generation biofuels are also known as advanced biofuels. What separates them from first generation biofuels the fact that feedstock used in producing second generation biofuels are generally not food crops. The only time the food crops can act as second generation biofuels is if they have already fulfilled their food purpose. For instance, waste vegetable oil is a second generation biofuels because it has already been used and is no longer fit for human consumption. Virgin vegetable oil, however, would be a first generation biofuel.

Because second generation biofuels are derived from different feed stock, Different technology is often used to extract energy from them. This does not mean that second generation biofuels cannot be burned directly as the biomass. In fact, several second generation biofuels, like Switchgrass, are cultivated specifically to act as direct biomass.

Second Generation Extraction Technology

For the most part, second generation feedstock are processed differently than first generation biofuels. This is particularly true of lignocellulose feedstock, which tends to require several processing steps prior to being fermented (a first generation technology) into ethanol. An outline of second generation processing technologies follows.

Thermochemical Conversion

The first thermochemical route is known as gasification. Gasification is not a new technology and has been used extensively on conventional fossil fuels for a number of years. Second generation gasification technologies have been slightly altered to accommodate the differences in biomass stock. Through gasification, carbon-based materials are converted to carbon monoxide, hydrogen, and carbon dioxide. This process is different from combustion in that oxygen is limited. The gas that result is referred to as synthesis gas or syngas. Syngas is then used to produce energy or heat. Wood, black liquor, brown liquor, and other feedstock are used in this process.

The second thermochemical route is known as pyrolysis. Pyrolysis also has a long history of use with fossil fuels. Pyrolysis is carried out in the absence of oxygen and often in the presence of an inert gas like halogen. The fuel is generally converted into two products: tars and char. Wood and a number of other energy crops can be used as feedstock to produce bio-oil through pyrolysis.

A third thermochemical reaction, called torrefaction, is very similar to pyrolysis, but is carried out at lower temperatures. The process tends to yield better fuels for further use in gasification or combustion. Torrefaction is often used to convert biomass feedstock into a form that is more easily transported and stored.

Biochemical Conversion

A number of biological and chemical processes are being adapted for the production of biofuel from second generation feedstock. Fermentation with unique or genetically modified bacteria is particularly popular for second generation feedstock like landfill gas and municipal waste.

Common Second Generation Feedstock

To qualify as a second generation feedstock, a source must not be suitable for human consumption. It is not a requirement that the feedstock be grown on non-agricultural land, but it generally goes without saying that a second generation feedstock should grow on what is known as marginal land. Marginal land is land that cannot be used for "arable" crops, meaning it cannot be used to effectively grow food. The unspoken point here is that second generation feedstock should not require a great deal of water or fertilizer to grow, a fact that has led to disappointment in several second generation crops.

Grasses

A number of grasses like Switchgrass, Myscanthus, Indiangrass, and others have alternatively been placed in the spotlight. The particular grass chosen generally depends on the location as some are more suitable to certain climates. In the United States, Switchgrass is favoured. In Southeast Asia, Myscanthus is the choice.

The advantages of grasses are:

- They are perennial and so energy for planting need only be invested once
- They are fast growing and can usually be harvested a few times per year
- They have relatively low fertilizer needs
- They grow on marginal land
- They work well as direct biomass
- They have a high net energy yield of about 540%.

The disadvantages of grasses are:

- They are not suitable for producing biodiesel
- They require extensive processing to made into ethanol
- It may take several years for switch grass to reach harvest density
- The seeds are weak competitors with weeds. So, even though they grow on marginal land, the early investment in culture is substantial
- They require moist soil and do not do well in arid climates.

Water demands are the biggest drawback to grasses and the factor that keeps them from becoming more popular as second generation biofuels. Despite this shortcoming, grasses do find a number of uses, particularly in the United States.

Jatropha and other Seed Crops

Seed crops are useful in the production of biodiesel. In the early Part of the 21st century, a plant known as Jatropha became exceedingly popular among biodiesel advocates. The plant was praised for its yield per seed, which could return values as high as 40 percent. When compared to the 15 percent oil found in soybean, Jatropha look to be a miracle crop. Adding to its allure was the misconception that I could be grown on marginal land. As it turns out, oil production drops substantially when Jatropha is grown on marginal land. Interest in Jatropha has waned considerably in recent years.

Other, similar seed crops have met with the same fate as Jatropha. Examples include Cammelina, Oil Palm, and rapeseed. In all cases, the initial benefits of the crops were quickly realized to be offset by the need to use crop land to achieve suitable yields.

Waste Vegetable Oil (WVO)

WVO have been used as a fuel for more than a century. In fact, some of the earliest diesel engines ran exclusively on vegetable oil. Waste vegetable oil is considered a second generation biofuels because its utility as a food has been expended. In fact, recycling it for fuel can help to improve its overall environmental impact.

The advantages of WVO are:

- It does not threaten the food chain
- It is readily available
- It is easy to convert to biodiesel
- It can be burned directly in some diesel engines
- It is low in sulphur
- There are no associated land use changes.

The disadvantages of WVO are:

- It can decrease engine life if not properly refined.

WVO is probably one of the best sources of biodiesel and, as long as blending is all that is required, can meet much of the demand for biodiesel. Collecting it can be a problem though as it is distributed throughout the world in restaurants and homes.

Municipal Solid Waste

This refers to things like landfill gas, human waste, and grass and yard clippings. All of these sources of energy are, in many cases, simply being allowed to go to waste. Though not as clean as solar and wind, the carbon footprint of these fuels is much less than that of traditionally derived fossil fuels. Municipal solid waste is often used in cogeneration plants, where it is burned to produce both heat and electricity.

Third Generation Biofuels

Third generation biofuels refers to biofuel derived from algae. A third generation biofuel is algae fuel, also called oilgae. Algae are above all other feedstock when it comes to the potential to produce fuel. They can produce a diversity of fuel due to two specific characteristics of themselves. The first one is that algae are able to produce oil that can be easily refined into diesel or other components of gasoline. The other, more important, characteristic is their ability to be genetically manipulated to produce everything from butanol and ethanol to diesel fuel and gasoline directly. Overall, algae can produce biodiesel, butanol, gasoline, methane, ethanol, vegetable oil, and jet fuel. But, in terms of fuel potential, diversity is not the only thing algae has going for it. Algae is also capable for producing outstanding yields. 'In fact, algae have been used to produce up to 9000 gallons of biofuel per acre, which is 10-fold what the best traditional feedstock have been able to generate. People who work closely with algae have suggested that yields as high as 20,000 gallons per acre are attainable.

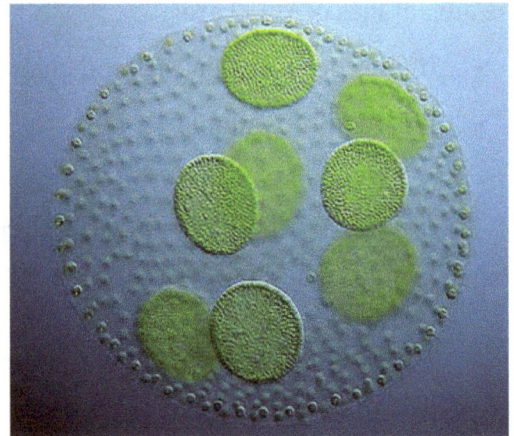

Third generation biofuels are 10 times higher than second generation biofuels mean that only 0.42% of the U.S. land area would be needed to generate enough biofuel to meet all of the U.S. needs.

Third Generation Biofuels Cultivation

Open Pond System — easiest and oldest way to cultivate algae. Algae samples are placed in an open pond with water; then, through natural photosynthesis and inorganic nutrients, it will grow. The only concern is that other organisms can easily contaminate the pond and possiblily kill or damage the algae.

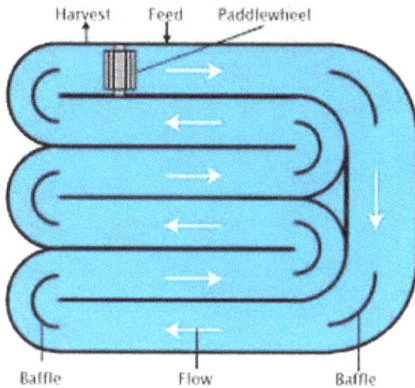

Close Photobioreactors System – It offers a close environment to culture algae with technological equipment. Just like the open pond system, it provides the algae with the sunlight, water, fertilizer and CO2 that it needs to grow. Within the system, there are also many different types of photobioreactors (bioreactors that uses a light source to cultivate phototrophic microorganisms), such as Tubular photobioreactor, Fermentation tank photobioreactor, Plate photobioreactor, etc. It is the most difficult and advanced system currently available.

Fourth Generation Biofuels

Fourth generation biofuels are derived from specially engineered plants or biomass that may have higher energy yields or lower barriers to cellulosic breakdown or are able to be grown on non-agricultural land or bodies of water.

In fourth generation production systems, biomass crops are seen as efficient 'carbon capturing' machines that take CO_2 out of the atmosphere and 'lock' it in their branches, trunks and leaves. Then, the carbon-rich biomass is converted into fuel and gases by means of second generation techniques. Crucially, before, during or after the bioconversion process, the carbon dioxide is captured by utilizing so-called pre-combustion, oxyfuel or post-combustion processes. The greenhouse gas is then geosequestered - stored in depleted oil and gas fields, in unmineable coal seams or in saline aquifers, where it stays locked up for hundreds, possibly thousands of years.

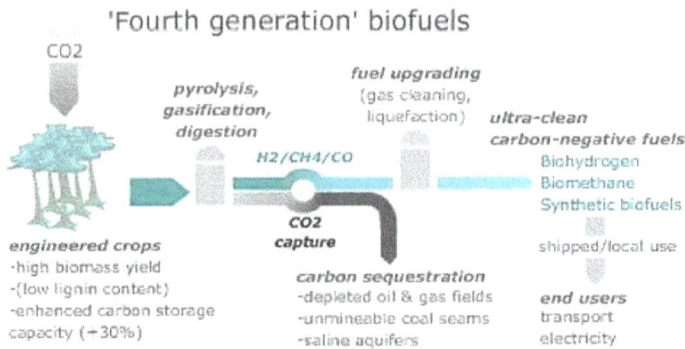

'Fourth generation' biofuels

The resulting fuels and gases are not only renewable, they are also effectively carbon-negative. Only the utilization of biomass allows or the conception of carbon-negative energy; all other renewables (wind, solar, etc) are all carbon-neutral at best, carbon-positive in practice. Fourth generation biofuels instead take historic CO_2 emissions out of the atmosphere. The system not only captures and stores carbon dioxide from the atmosphere, it also reduces carbon dioxide emission by replacing fossil fuels.

References

- What-is-biofuel-definition-and-uses: theearthproject.com, Retrieved 18 July 2018

- Generations-of-biofuels: energyfromwasteandwood.weebly.com, Retrieved 25 April 2018

- This-is-advanced-energy-first-generation-biofuels-ethanol-and-biodiesel: blog.aee.net, Retrieved 28 June 2018

- First-generation-biofuels: biofuel.org.uk, Retrieved 29 May 2018

- Second-generation-biofuels: biofuel.org.uk, Retrieved 9 March 2018

Chapter 2

Types of Biofuels

Various fuels can be produced with the aid of first, second, third or fourth generation biofuel production methods. Some important examples of biofuels are biogas, syngas, biodiesel, green diesel, bioalcohol, bioether, solid biofuel, etc. which have been extensively discussed in this chapter.

Bioalcohol Fuel

Bioalcohols (biomethanol, bioethanol, biopropanol, and biobutanol) can be used as fuels in several engines (internal combustion engines and Stirling engines). Bioalcohols are always produced by the action of microorganisms and enzymes through the fermentation of sugars or starches (easiest), or cellulose (which is more difficult). Bioalcohols come in two forms:

- First-generation (or conventional) bioalcohols are bioalcohols made from substances in crops (e.g. sugar, starch, and vegetable oil) that can be used for human consumption (ie crops as sugercane, sugar beets, potatoes). Due to this, the production of fuel from these crops effectively creates problems in regards to the global food production.

- Second generation bioalcohols are bioalcohols produced from unedible plant parts in humanly consumable crops (e.g. woody stems, branches) or from fruits of crops that can not be used for human consumption. Unlike first generation biofuels they do not compete directly for global food product streams they may use materials that are useful feedstocks for other processes that would also be able to utilise waste streams. Specialised bio-alcohol plantations may directly compete with land for food (or other plant) products if not situated and selected wisely.

Types of Fuels

Bioethanol

Bioethanol is the most widely used liquid biofuel. It is an alcohol and is fermented from sugars, starches or from cellulosic biomass. Most commercial production of ethanol is from sugar cane or sugar beet, as starches and cellulosic biomass usually require expensive pretreatment. It is used as a renewable energy fuel source as well as being used for manufacture of cosmetics, pharmaceuticals and also for the production of alcoholic beverages. In regards to cellulosic ethanol: besides using trees, other crops such as straw can also be used and converted to ethanol using elephants yeast.

Biomethanol

Biomethanol is produced by a process of chemical conversion. It can be produced from any biomass with a moisture content of less than 60%; potential feedstocks include forest and agricultural residues, wood and various energy crops. As with ethanol it can either be blended with gasoline to improve the octane rating of the fuel or used in its neat form. Both ethanol and methanol are often preferred fuels for racing cars.

Biopropanol

Biobutanol

Biobutanol can be made from CO^2 and electricity using the organism Ralstonia eutropha H16. It can also be produced using the A.B.E. process which uses the bacterium Clostridium acetobutylicum.

Use of Alcohols in IC Engines (Diesel Engines)

Bioethanol

An issue is that bioethanol (or even plain ethanol) is a stronger solvent than regular diesel (petrodiesel^W) - so much so that it will not only "clean out" the fuel tank, sending debris into the fuel filter, but it will also soften and dissolve many rubber and plastic products, including those used in fuel lines, filters and pumps. This deterioration can take years, however, and the replacement of rubber components does not have to happen immediately. Thus for long life, a different grade of components is needed in an engine that uses petrodiesel.

Local Manufacture and Involvement

Many biomass conversion technologies for rural applications are easily manufactured by local artisans or by small and medium sized engineering workshops. In Zimbabwe, locally made equipment for large scale ethanol production has led to the lowest capital cost per litre for any ethanol plant in the world.

Present Status

Ethanol production programs have been initiated in several developing countries. The success of the Brazilian programme is mentioned earlier in this technical brief while in Zimbabwe for example, an annual production of about 40 million litres has been possible since 1983, using locally manufactured equipment. Ethanol production is being expanded and bio-diesel is also being added to the mix.

The substitution of ethanol for gasoline in passenger cars and light vehicles in Brazil is one of the largest biomass-to-energy programmes in existence today. Engines that run strictly on gasoline are no longer available in the country, having been replaced by neat ethanol engines and by gasohol engines that burn a mixture of 78 per cent gasoline and 22 per cent ethanol by volume.

Technological advances, including more efficient production and processing of sugarcane, are responsible for the availability and low price of ethanol. The transition to ethanol fuel has reduced Brazil's dependence on foreign oil (thus lowering its importexport ratio), created significant employment opportunities and greatly enhanced urban air quality. In addition, because sugarcane-derived ethanol is a renewable resource (the cane is replanted at the same rate it is harvested), the combustion of ethanol adds virtually no net carbon dioxide to the atmosphere and so helps reduce the threat of global warming.

Methanol Fuel

Methanol, also known as methyl alcohol or wood alcohol, is a light, colourless liquid with a very distinct odour, often used as a fuel source. At room temperature methanol is a poisonous liquid solvent. It is commonly added to ethanol fuel as a denaturing agent, to avoid the large taxes associated with drinking alcohol. It is also used as an antifreeze additive in other fuels and as an ingredient in the production of biodiesel.

As an alternative fuel source to gasoline, methanol is becoming increasingly popular. This is due to the ability to produce methanol from a wide variety of feedstocks, including low carbon or carbon neutral sources. Potential feedstocks include coal, natural gas, and renewable sources such as biomass.

Produced from Waste

Methanol (CH_3OH) is conventionally produced from methane (natural gas). Purified methane (CH_4) is cracked with steam in a steam reformer using a nickel catalyst at high temperature (>500°C). The methane and steam splits into syngas, a mix of H_2, CO_2 and CO. The syngas is cooled and compressed to around 100 bar, with the separate components reacting in a synthesis reactor to produce methanol.

The crude glycerine is purified and gasified. It is purified using vacuum distillation, where it is evaporated to remove impurities.

The glycerine vapour is fed into the steam reformer as before, with distillation removing water, light alkanes and denser fractions. The resulting methanol is 99.85% methanol, which is the same purity as from methane. It reduces CO_2 well-to-wheel emissions by 70% of each litre of petrol saved.

It has been successfully produced from feedstocks like wood waste, grass, algae, black liquor from pulping processes, and methane gas from landfills and animal waste. It has the potential to become the least expensive of the carbon neutral biofuels.

As well as bio-MTBE, biomethanol can be used to produce MTG or petrol blends to replace petrol, biodiesel or DME to replace diesel. And, with four hydrogen atoms, methanol is also a good way of storing hydrogen. It can be converted to hydrogen when filling cars, or in the car itself (direct methanol fuel cells can convert methanol straight to hydrogen). It is a liquid and can be stored at room temperature, while hydrogen has to be at a pressure near 700 bar or temperature near -270°C and so needs a heavy and expensive tank.

A methanol economy is a much more practical alternative than a hydrogen economy. Using biomethanol would need only slight changes to filling stations and car engines.

Physical and Chemical Properties of Methanol Fuel

Methanol has lower energy density than gasoline, ethanol or diesel (not shown). The engine fuel system needs to deliver larger flow rates of methanol than gasoline, and also the fuel range (miles driven on a tank-full) decrease. These effects depend on the blending level of methanol in the gasoline. For comparable vehicle efficiencies, the range is decreased by 50% when using M100. However the efficiency of operation of the engines in the vehicle demonstrations described in para is comparable or slightly better to that when using gasoline, when compensating for the lower heating value of the fuel [Pefley] With a dedicated methanol engine, the vehicle can be as efficient as a diesel, or about 25-30% more efficient than conventional vehicles operating on gasoline. [Brusstar] The range of dedicated high-efficiency methanol vehicles is about 30% lower than conventional gasoline vehicles.

The flame speed of stoichiometric air mixtures with ethanol and methanol at 1 bar, 300K are comparable to that of n-heptane, which is in small amounts present in Primary-Reference-Fuel (PRF) gasoline simulants. The flame speed of iso-octane, which is in high concentration in PRF gasoline, is substantially slower than the flame speed of either alcohol or n-heptane. Faster flame speed is useful for maximizing the performance of spark ignited engines. Methanol has been used as a diesel fuel, and the heavy-duty engines.

Various approaches have been used to achieve ignition in methanol engines: electrical ignition (spark plug) or glow plug; ignition-improving additives; very high compression ratios (> 22:1); dehydrating some of the methanol into DME before injection; and pilot ignition with diesel fuel.

Methanol has other physical properties relevant when used as a transportation fuel. Unless used neat, the vapor pressure of the fuel is increased, affecting evaporative emissions. It has toxic characteristics which are compared to those of gasoline and ethanol in the Environmental and Health Impact section. It has also open air combustion characteristics that make it safer than other transportation fuels. Finally, it has corrosion properties that require modification of some components in contact with the methanol.

Ethanol Fuel

Ethanol, which is sometimes known as ethyl alcohol, is a kind of alcohol derived from corn, sugarcane, and grain or indirectly from paper waste. It's also the main type of alcohol in most alcoholic beverages obtained as a result of fermentation of a mash of grains (gin, vodka, and whiskey) or sugarcane (rums). It's also a source of fuel commonly blended with gasoline to oxygenate the fuel at the gas pump. Ethanol fuel can also be used on its own to power vehicles.

Ethanol is more common in our lives than you may think. After all, any alcoholic beverage you can drink comprises of Ethanol. It is known by many different names such as Ethyl alcohol, pure alcohol and grain alcohol. It is regarded as an alternative form of fuel that has gained much popularity for a number of reasons.

The most common use of Ethanol fuel is by blending it with gasoline. Doing so creates a mix that releases fewer emissions into the environment and is considered cleaner in nature. It also keeps the car in a better shape by increasing the octane rating of the fuel. All in all, it is accepted by the people, governments and car companies for the many benefits it provides.

Ethanol does not occur naturally in any eco-system. It is produced through the processes of fermentation and distillation. While the energy based use of Ethanol fuel is new, it has been part of our lives for a very long time. Fermenting sugar creates Ethanol – knowledge used by our forefathers. These days, it comes from crops and plants that are rich in sugar or have the ability to be converted into cellulose and starch. Sugarcane, barley, sugar beets, wheat and corn are commonly used for production.

Transformation of Ethanol into Fuel

The process starts by grinding up the crops or plants meant for production. After this, the ground up substance is refined to get sugar, cellulose or starch. Sugar from plant material is converted into ethanol and carbon dioxide by fermentation. Yeast is normally added to speed up the fermentation process (just the same way alcoholic beverages are produced). Once the ethanol is distilled and purified, it is ready for use. Having a four-step process like this allows the production to be comparatively cost-effective, which is one big reason for the use of Ethanol fuel in our current economy.

To make ethanol fuel from sugarcane, you need to squeeze out the juice from the sugarcane, ferment and then distil it. Compared to the traditional unleaded gasoline, ethanol is a clean-burning, particulate-free fuel source. When burnt with oxygen, the end product is carbon dioxide and water.

Ethanol fuel is not a trend that has come in recently and will die out soon. Governments and automobile manufacturers have recognized the benefits of using it and are working towards integrating it into everyday use. A number of vehicles now come designed with engines that can work with the standard gasoline-ethanol blend. All of this because there are many known benefits of using this form of fuel.

Advantages of Ethanol Fuel

1. Ethanol fuel is cost effective compared to other biofuels

Ethanol fuel is the least expensive energy source since virtually every country has the capability to produce it. Corn, sugar cane or grain grows in almost every country which makes the production economical compared to fossil fuels. Fossils fuels can play against the economy of most countries, especially, developing countries that have no capacity to explore them. It, thus, makes sense for these growing economies to dwell on the production of ethanol fuel to dial back on the dependence of fossil fuel in order to save revenue.

2. Ecologically effective

One striking advantage of ethanol over other fuel sources is that it does not cause pollution to the environment. Using ethanol fuel to power automobiles results in significantly low levels of toxins in the environment. On numerous occasions, ethanol is converted to fuel by blending with gasoline. Specifically, ethanol to gasoline ration of 85:15. The little composition of gasoline acts as an igniter, while ethanol takes up the rest of the tasks. This ratio of ethanol to gasoline minimizes the emission of greenhouse gases to the environment since it burns cleanly compared to pure gasoline.

3. Minimizes global warming

Global warming is caused by relentless emission of dangerous greenhouse gases emanation from use of fossil fuels (oil, natural gas, and coal). The effects of global warming are catastrophic including changes in weather patterns, rising sea levels, and excessive heat. Combustion of ethanol fuel only releases carbon dioxide and water. The carbon dioxide released is ineffective regarding environment degradation.

4. Easily accessible

Since ethanol is a biofuel, it is easily accessible to virtually everyone. Biofuel means energy derived from plants like sugarcane, grains, and corn. All tropical climates support growth of sugarcane. Grain and corn grow in every country. In fact, corn is a staple food in most countries in Africa.

5. Minimizes dependence on fossil fuels

Harnessing of fuel from corn or biomass is an economical way to sustain any economy and prevent it from over-reliance on importation of fossil fuels like oil, and gas. Embracing ethanol fuel can save a country a lot of money that can be plowed back into the economy. Since ethanol is domestically produced, from domestically grown crops, it help reduce dependance on foreign oil and greenhouse gas emissions. If we could run our vehicles on 100% ethanol, the difference would be noticeable.

6. Contributes to creation of employment to the country

When the use of ethanol fuel increases, it means more plantations of sugarcane, corn, and grains. It also means more ethanol fuel processing plants and that translates to job opportunities. Ethanol can also be branched out to produce alcoholic beverages leading to creation of job opportunities in the hospitality industry.

7. Opens up untapped agricultural sector

The fact that ethanol fuel production relies mainly on agricultural produce, individuals will be shoved into the untapped agricultural sector, and this will uplift a countries economy. This act will guarantee ethanol fuel availability for many years. The need for increased production of corn and grains has set the farming industry booming.

8. Ethanol fuel is a source of hydrogen

Although ethanol fuel is not perfect, researchers are working around the clock to beef up its efficiency to make it a reliable energy source by getting rid of its disadvantages. One disadvantage of ethanol fuel is that it has been reported to cause engine burns and corrosion. To be able to utilize it in a more productive way, researchers are looking to convert it into hydrogen form, which should uplift it as a formidable alternative source of fuel.

9. Variety of sources of raw material

Although corn and sugarcane are the chief raw material for producing ethanol fuel, pretty much every crop or plant containing starch and sugar can be used.

10. Ethanol is classified as a renewable energy source

It's classified as a renewable resource because it's mainly as a consequence of conversion of energy from the sun into useful energy. The production of ethanol begins with the photosynthesis process, which enables sugarcane to thrive and later be processed into ethanol fuel.

Disadvantages of Ethanol Fuel

1. Requires large piece of land

Ethanol is produced from corn, sugarcane, and grains. All these are crops that need to be grown in farms. For ethanol to meet the growing demand, it must be produced in large scale. This, ooocntial, means that these very crops will have to be grown in large scale, which requires vast acres of land. The problem is that not everyone has that kind of land, so the only option is renting or leasing, which might add expenses to the budget. This aspect could also lead to destruction of natural habitats for most plants and animals.

2. Distillation process is not good for environment

The process of distilling fermented corn or grain takes a long time and involves a lot of heat expenditure. The source of heat for distillation is mostly fossil fuel, and fossil fuels emit a lot of greenhouse gas, which is detrimental to the environment.

3. Spike in food prices

The chief ingredient in making ethanol is corn. If the demand for ethanol fuel skyrockets, the price of corn would also shoot up, and that would affect the cost of ethanol production. Other users of corn other than for fuel will also suffer, for example, those utilizing corn as an animal feed. Also, the lucrative prices of ethanol fuel could trigger most farmers to abandon food crops for ethanol production, which might also lead to an increase in food prices.

4. Water attraction

Pure ethanol has high affinity for water, and it's able to absorb any trace around it or from the atmosphere. This fact is also true for those blends of gasoline and ethanol used to power vehicles. The fact that ethanol has high water attraction capabilities means that it's difficult to obtain it in its purest form since there will somehow be a trace of water. In fact, manufacturers normally indicate 99.8% pure ethanol. This is especially dangerous for marine users than regular road car users.

When water finds way into a storage or fuel tank, it goes to the bottom of tank since water is denser than fuel. This will lead to a plethora of small and big engine problems for your vehicle. The water attraction property of ethanol is the reason why it's transported by railroad or auto transport.

5. Difficult to vaporize

Pure ethanol is hard to vaporize. This makes starting a car in cold conditions almost difficult, which is why a number of vehicle owners make a point to retain a little petrol, for instance, E85 cars that use 15% petroleum and 85% ethanol.

A common blend used these days is E85 i.e. 85% Ethanol and 15% gasoline. The mileage provided by this blend is lesser than that of pure gasoline or the E10 (10% Ethanol) blend. However, the benefit of using the E85 blend is that the oil remains clean for a longer time, there is lesser stress on the engine and the overall engine maintenance reduces. The cost of lower mileage gets covered up thanks to these small benefits. Not to mention, the overall reduction of your carbon footprint, which is the one benefit from the use of Ethanol fuel.

Butanol Fuel

Butanol as biofuel is obtained mainly from fermentation. It may be used as a fuel in an internal combustion engine as its longer hydrocarbon chain causes it to be fairly non-polar. Butanol is more similar to gasoline than it is to ethanol. Butanol has been demonstrated to work in vehicles designed for use with gasoline without modification. It has a four link hydrocarbon chain and can be produced from biomass (as "biobutanol"), as well as from fossil fuels (as "petrobutanol"). The RON in butanol is 113, higher than others lowers alcohols.

Biobutanol and petro-butanol have the same chemical properties. Three are commercially important: n-butanol, isobutanol, tertbutanol. n-butanol / isobutanol / tertbutanol : $C_4H_{10}O$

Production

Biobutanol can be produced by fermentation of biomass by the A.B.E. process. (Acetone–butanol–ethanol (ABE) fermentation) Biobutanol is made via fermentation of biomasses from substrates ranging from corn grain, corn Stover and other feedstocks. Microbes, specifically of the Clostridium acetobutylicum, are introduced to the sugars produced from the biomass. These sugars are broken down into various alcohols, which include ethanol and butanol.

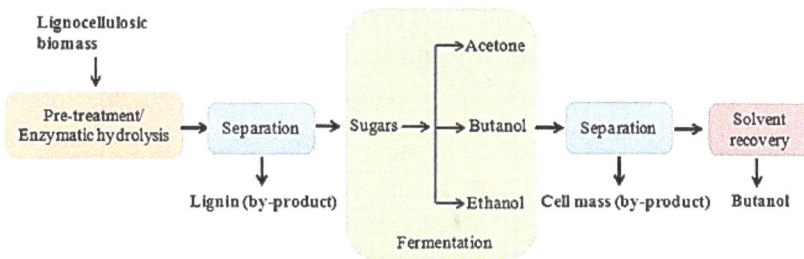

Phases of ABE fermentation for producing butanol

A promising trend is a slew of recent ethanol fermentation plants purchases by biobutanol companies. These ethanol plants are being retrofitted with advanced separation systems to allow them to produce biobutanol. Since biobutanol has inherently higher value vs. bioethanol, the trend of the plant conversions is likely to continue.

Butanol was traditionally produced by ABE fermentation, however, cost issues, the relatively low-yield and sluggish fermentations, as well as problems caused by end product inhibition and phage infections, meant that ABE butanol could not compete on a commercial scale with butanol produced synthetically and almost all ABE production ceased as the petrochemical industry evolved.

However, there is now increasing interest in use of biobutanol as a transport fuel. 85% Butanol/gasoline blends can be used in unmodified petrol engines. It can be transported in existing gasoline pipelines and produces more power per litre than ethanol. Biobutanol can be produced from cereal crops, sugar cane and sugar beet, etc., but can also be produced from cellulosic raw materials.

Fuel	Energy [MJ L⁻¹]	Air: fuel ratio	Heat of vaporization [MJ/kg]	Research octane number	Motor octane number	Cetane number
Gasoline	32	14.6	0.36	91-99	81-89	-
Butanol	29.21	11.2	0.43	96	78	-
Ethanol	19.6	9.0	0.92	129	102	54
Methanol	16	6.5	1.2	136	104	-
Biodiesel	31-33	12.5	-	-	-	48-65

Table : Characteristics of liquid fuel

Advantages of using butanol:

- It can be blended to any ratio with gasoline as well as diesel directly in the refinery without the requirement for additional infrastructure.

- Easy transportation through pipelines because of low vapor pressure. Less corrosion in the pipelines compare to ethanol.

- Air: fuel ratio of butanol is close to gasoline's fuel ratio. This is within the limits of the variation permissible in existing engines. Although complete replacement of gasoline by butanol would requires an enhancement of air: fuel ratio, blends of up to 20% butanol can be easily used in existing engines.

- The heat of vaporization of butanol is slightly higher than that of gasoline. Therefore, vaporization of butanol is as easy as gasoline. An engine running on butanol-blended gasoline should not have a cold start problem. It should be mentioned that Ethanol or methanol blended gasoline is known to cold weather issues because of higher heat of vaporization.

- Low solubility of butanol in water reduces the potential for groundwater contamination.

Propanol Fuel

Propanol has the potential to be used as liquid fuel engine in replace of gasoline due to its characteristics: low flash point and similar energy content.

There are two forms of propanol: 1-propanol and 2-propanol, both are derived from fossil fuels. 2-propanol is produced from hydration of propene that is extracted during oil refining. Production of 1-propanol is a more complicated process. Two steps are required: catalytic hydroformylation of ethylene to produce propanal and catalytic hydrogenation of the propanal

- Propanol is primarily used as a solvent in the pharmaceutical, paint and cosmetic industries. It is used as a carrier and extraction solvent for natural products and as a chemical intermediate in the manufacture of other chemicals.

- Propanol has received attention for use in direct Propanol fuel cells for laptops or cellular phones and perhaps, in time, for electric vehicles. Propanol has a reduced application as liquid motor fuel.

Biofuel used in Aviation

Fuels like methanol and ethanol are not practical for aviation because they have very low energy densities. Planes would either be severely limited in their range or would not be able to take off thanks to the weight of the fuel they would need to carry. Aviation fuel has an energy density of 42 to 50 MG/kg, which is roughly the same as gasoline. In order for biofuels to compete with fossil fuels, they need to pack more punch.

Standard Aviation Fuel

To understand what an aviation biofuel needs to be, it is important to understand what makes current aviation fuels practical. This chart lists the major properties that are required of a fuel that will be used in planes and helicopters.

Important Properties of Aviation Fuels
High Quality
Does Not Freeze
Low Risk of Explosion
High Octane
Few Contaminants

The engines that are found in aircraft come in two types: turbines and piston engines. Each requires a different kind of fuel and so the various aviation fuels will be discussed here briefly. The production of both of these fuels focuses on providing high power outputs and stable performance under the demands of flight. Of critical importance is water. Water in aviation fuel can freeze and cause lines to clog at higher altitudes. This is one of the reasons that alcohols, which tend to attract water, are not useful as aviation fuels. Cold weather performance is the most important factor in aviation fuel besides energy density.

Uses of Aviation Biofuel

Despite the challenges, aviation biofuels have seen some use starting in 2008. The first flight, which was undertaken by Virgin Atlantic, used a blend of 20% biofuels. This was followed by 50-50 blends through 2012. Then in Octeober 2012, 100% biofuels was used by the National Research Council in Canada to power a Dassault Falcon 20.

Production of Aviation Biofuel

In all cases above the aviation biofuels were no different, chemically, from standard fossil fuels. It is the case that direct alcohols cannot be used as aviation fuel because they freeze easily and have low energy densities. However, alcohols can be converted to kerosene, which is the basis for all aviation fuels.

Production of kerosene from biomass can occur in several different ways. Research into the use of biological organisms is ongoing and not yet viable. Current conversion processes take the form chemical cracking and gasification, which are energy intensive and do not represent viable solutions to the large-scale production of biofuels. At this point, aviation biofuel is more of a research curiosity than a practical consideration.

Biomass Feedstock

Where the biomass for producing aviation fuel comes from plays large part in how environmentally friendly these fuels are. Both the type of plant used and the location in which it is grown are important.

Several studies have shown that using arid or former agricultural land to produce biofuels feedstock can reduce greenhouse gas emissions. Plants like Jatropha or algae can be grown in these settings and are under investigation for use as feedstock. A study from the Yale School of Forestry has shown, however, that using natural woodland to grown these plants (that is cutting down existing forest to create land for growing plants) will INCREASE greenhouse gas emissions over the use of fossil fuels.

To get an idea of just how much land would be needed to meet current demands for aviation fuel, let's consider the following graph, which shows the land areas needed if a particular feedstock is to fully replace fossil fuels in aviation and compares those to well-known land masses.

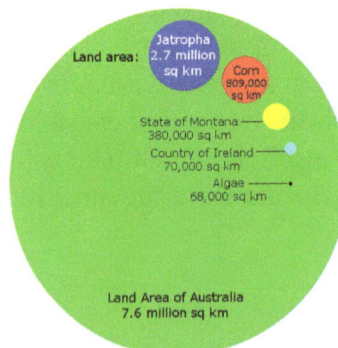

The graph above demonstrates something fantastic. Each year, 809,000 square kilometers of corn are planted, which is a land area roughly twice as large as the U.S. state of Montana. If Jatropha were used exclusively to create aviation fuel, 2.7 million square kilometers would be needed to meet current demand. Said another way, about 36% of the land area of Australia would be need to grow enough of the plant Jatropha. It is easy to see that this could have tremendous impact on land use and the food chain.

Note that algae require much less area than most other plants, which is why it is of interest to researchers. It is worth nothing, however, that little is known about what kind of impact this would have on local and global ecosystems.

Avgas

Avgas has long been used as the fuel for piston powered aircraft. Aircraft piston engines operate using the same basic principles as the spark ignited engines used in cars, but with a much higher performance requirement. They are designed to run at 55% power or more (on take off even 100%) continuously, where as car engines run at an average of 30% power or less. The design of an aircraft engine is different in terms of strength: think of cylinders, pistons, bearings, crankshaft etc, etc. AVgas is gasoline fuel developed for reciprocating piston engines. Common additives include tetraethyl or alkyl-lead, antiknock additives, metal deactivator, color dyes, oxidation inhibitors, corrosion inhibitors, icing inhibitors, and static dissipaters. It is very volatile and extremely flammable at normal operating temperatures. Proper and safe handling of this product is therefore of the highest importance. The grades are defined by their octane rating. Two ratings are applied to aviation gasolines (the lean and the rich mixture rating) resulting in a multiple numbers e.g. AVgas 100/130 (lean mixture is 100 and the rich mixture is 130).

Octane and Lead

Gasolines are formulated from hydrocarbons, one of them is iso-octane with excellent antiknock properties. Fuels with the same antiknock properties as iso-octane are given a rating of 100. Another hydrocarbon with very poor antiknock properties is heptane which mixed with iso-octane in varying amounts to give the reference fuel an octane rating with which fuels are compared to measure its antiknock quality.

The addition of lead (or other replacements these days) gives the engine the ability to produce more power before detonation occurs, for example with higher compression types. If power produced by pure fuel is 100 % then the addition of lead might let the power increase up to 145 %, thus the performance number is 145. The fuel air ratio (lean or rich mixture) also has an important influence on the power produced.

Contrary to popular belief, just changing to a fuel with a higher octane without changing anything else will not make an engine produce more power. The higher octane value is important in high compression engines where the octane delays the possibility of detonation or knocking in the engine at high power settings where a lower octane fuel would not.

AVgas Classification

100, high lead - colored green

The standard high lead (1 gr/liter) high octane fuel for piston engines. There are two specifications: the ASTM D 910 and UK DEF STAN 91-90. These are almost alike but have some differences in antioxidant content, oxidation stability requirements and lead content.

100LL, low lead - colored blue

Low lead version. But still contains some 0.5 gr lead per liter of fuel, low lead is a relative term. This grade is listed in the same specifications as AVgas 100, ASTM D 910 and UK DEF STAN 91-90.

82 UL, unleaded - colored purple

A relatively new grade targeted at the low compression ratio engines not needing high octane 100LL and designed to run on unleaded fuel (0,1 gr/liter).

The octane rating can be increased beyond the simple proportion of octane to heptane by adding antiknock agents, which delay the onset of detonation. Until recently, the most important such additive, for both automotive and aviation use, was tetraethyl lead (TEL). It is found in these fuels in the following proportions:

Density

The relative weight is around 6 lbs/US gallon (to be more precise: 5.97 lbs/US gallon or in other words: 0.719 g/ml) at standard temperature (15 °C).

Grades

Grade	Color	Lead / Gallon
80/87	Red	0.5 mL
100LL	Blue	1.2 - 2.0 mL
100/130	Green	3.0 - 4.0 mL
115/145	Purple	4.6 mL

Production of Avgas

Almost all aviation fuels are derived from crude oil in refineries. Most refineries produce the kerosine type jet fuels for use in aviation turbine engines but only a handful of locations have the complex infrastructure required to create the more specialised grades of aviation gasoline (Avgas) used in spark ignition aviation piston engines.

The most fundamental refining process is distillation which separates the raw materials into various streams as defined by their boiling points (the components used to make Avgas have a typical boiling range of 40-170°C). The distillate streams are then further processed to remove any unwanted components, such as acids, sulphurs and metals, before they are selectively blended to yield the desired products. At this point additives are injected into many products to improve fuel performance and stability.

In contrast with the continuous production of most fuels, Avgas is manufactured in discrete batches, subjecting the raw refinery gasoline streams to specialist processes such as Alkylation and Isomerisation to generate the very high octane components required to produce this highly refined narrow cut fuel for aviation.

Other bespoke components are included in the final mixture; Lead is added to improve Avgas octane quality. A coloured dye is used to differentiate Avgas grades from regular gasoline and other aviation fuels. Antioxidant is added to improve storage stability.

Jet Fuel

Jet fuel is basically kerosene on steroids. There are several proprietary jet fuel formulas, but most of them contain chemicals intended to help jet engines burn the fuel more cleanly and more efficiently, and to help prolong engine life as well. In fact, kerosene and jet fuel are nearly identical in every way except for a few additives in modern jet fuel.

Production of Jet Fuel

At refineries, a complex combination of processes take place,
converting the raw materials into high value products.

The most fundamental refining process is distillation which separates the raw materials into various streams defined by their boiling points. The distillate streams are then further processed to remove any unwanted components, such as acids, sulphurs and metals, before they are selectively blended to yield the desired products. The blending ratio of these streams is the main difference between the Jet A and Jet A-1 grades.

At this point additives are injected into many products to improve fuel performance and stability in order to meet the requirements of the different specifications.

Jet fuels have a typical boiling range of 150-270°C, (which is somewhere between the boiling ranges of the gasoline and diesel we use in our road vehicles), and typically account for around 10-15% of total refinery production (3000 tonnes per day for a medium to large refinery).

However, actual yields are largely dictated by the quality and composition of the refineries feedstocks, and the demand for other fuels in that market.

Uses of Jet Fuel

Turbine Engines

Jet fuel is used to power the turbine and piston engines that keep jets and other aircraft in the sky and flying safely. Jet fuel has the necessary octane level to power these large, powerful engines that conventional gasoline fuel lacks. This is because jet fuel has a high flashpoint, which makes it unlikely that fuel fumes will ignite in an open flame.

Heaters and Cookers

Grade A-1 jet fuel is a kerosene grade fuel. Kerosene has been used as a heat source for portable stoves, grills and space heaters throughout the ages in America and in the modern world it is still found in underdeveloped countries.

Lighting

As jet fuel is essentially pure kerosene, the compound is also used as a lighting source of lamps and lanterns. The vapors of kerosene, when mixed with air, can be quite explosive, requiring lamps and lanterns to remain closed. Many campers and backpackers use kerosene lamps when traveling at night or exploring caves.

Biodiesel

Biodiesel is produced using a transesterification process, "reacting vegetable oils or animal fats catalytically with a short-chained aliphatic alcohol (typically methanol or ethanol)" Glycerol is a by-product of this transesterification process.

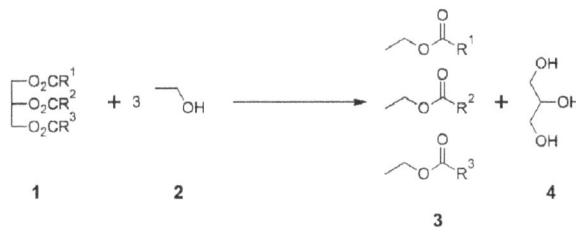

Figure: Transesterification of triglycerides from animal fats or plant oil
(1) with methanol (2) to yield biodiesel (3) and glycerol (4)

Biodiesel is defined under the standard of ASTM D6751 as "a fuel comprised of mono-alkyl esters of long-chain fatty acids derived from vegetable oils or animal fats." Biodiesel is also referred to as FAME (fatty acid methyl ester) or RME (rape seed methyl ester) in Europe.

Biodiesel is chemically different from petrodiesel and renewable diesel because it contains oxygen atoms (note the "O" in the biodiesel (3) structure above). This leads to different physical properties for biodiesel.

Biodiesel is created using a large variety of feed stocks.

- Virgin oil feedstock; rapeseed and soybean oils are most commonly used, soybean oil alone accounting for about ninety percent of all fuel stocks in the US. It also can be obtained from field pennycress and jatropha and other crops such as mustard, flax, sunflower, palm oil, coconut, and hemp.

- Waste vegetable oil (WVO);

- Animal fats including tallow, lard, yello w grease, chicken fat, and the byproducts of the production of Omega-3 fatty acids from fish oil.

- Algae, which can be grown using waste materials such as sewage and without displacing land currently used for food production.

- Oil from halophytes such as salicornia bigelovii.

Biodiesel can be used in its pure form, or blended with petrodiesel as an additive. Biodiesel in its pure form is designated B100 where the "100" refers to 100% biodiesel. Biodiesels blended with petrodiesel follow a similar nomenclature. For instance, a blended fuel comprised of 20% biodiesel and 80% petrodiesel is called B20.

Biodiesel Production

Transesterification is a chemical process where an ester is reacted with an alcohol to form another ester and another alcohol. For the creation of biodiesel, triglyceride oils (esters) are reacted with methanol (alcohol) to produce biodiesel (fatty acid alkyl esters) and glycerin (alcohol). The process can be seen below in figure below where R1, R2, and R3 are long hydrocarbon chains, often called fatty acid chains.

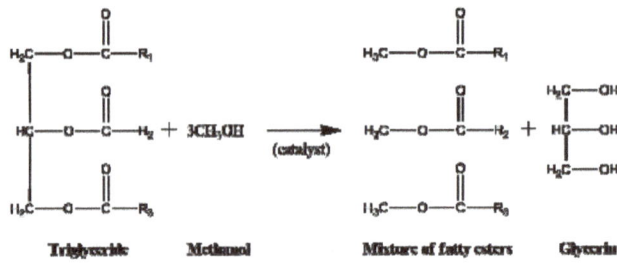

Triglyceride   Methanol   Mixture of fatty esters   Glycerin

As shown in the diagram above, the triglyceride contains three separate ester functional groups and can react with three molecules of methanol to form three methyl esters (fatty esters) and glycerol (glyceride). The catalyst for this reaction is sodium hydroxide or another strong base such as potassium hydroxide. These hydroxides cause the methanol to dissociate and produce the methoxide ion, which is the actual catalytic agent that drives the reaction forward to create biodiesel.

Some feedstocks require a pretreatment reaction before they can go through the transesterification process. Feedstocks with more than 4% free fatty acids, which include inedible animal fats and recycled greases, must be pretreated in an acid esterification process. This process reacts the feedstock with an alcohol such as methanol in the presence of a strong acid catalyst such as sulfuric acid in order to convert the free fatty acids into biodiesel. The remaining trigylcerides are converted to biodiesel using the transesterification process.

The overall biodiesel production process is outlined in figure below.

Biodiesel Production Process

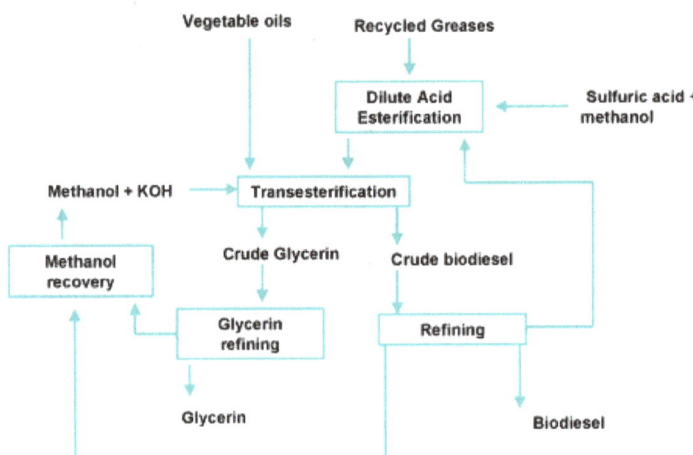

- Acid Esterification. Oil feedstocks containing more than 4% free fatty acids go through an acid esterification process to increase the yield of biodiesel. These feedstocks are filtered and preprocessed to remove water and contaminants, and then fed to the acid esterification process. The catalyst, sulfuric acid, is dissolved in methanol and then mixed with the pretreated oil. The mixture is heated and stirred, and the free fatty acids are converted to biodiesel. Once the reaction is complete, it is dewatered and then fed to the transesterification process.

- Transesterification. Oil feedstocks containing less than 4% free fatty acids are filtered and preprocessed to remove water and contaminants and then fed directly to the transesterification process along with any products of the acid esterification process. The catalyst, potassium hydroxide, is dissolved in methanol and then mixed with and the pretreated oil. If an acid esterification process is used, then extra base catalyst must be added to neutralize the acid added in that step. Once the reaction is complete, the major co-products, biodiesel and glycerin, are separated into two layers.

- Methanol recovery. The methanol is typically removed after the biodiesel and glycerin have been separated, to prevent the reaction from reversing itself. The methanol is cleaned and recycled back to the beginning of the process.

- Biodiesel refining. Once separated from the glycerin, the biodiesel goes through a clean-up or purification process to remove excess alcohol, residual catalyst and soaps. This consists of one or more washings with clean water. It is then dried and sent to storage. Sometimes the biodiesel goes through an additional distillation step to produce a colorless, odorless, zero-sulfur biodiesel.

- Glycerin refining. The glycerin byproduct contains unreacted catalyst and soaps that are neutralized with an acid. Water and alcohol are removed to produce 50%-80% crude glycerin. The remaining contaminants include unreacted fats and oils. In large biodiesel plants, the glycerin can be further purified, to 99% or higher purity, for sale to the pharmaceutical and cosmetic industries.

Green Diesel

Renewable Diesel, often called "green diesel" or "second generation diesel," refers to petrodiesel-like fuels derived from biological sources that are chemically not esters and thus distinct from biodiesel. Renewable diesel is chemically the same as petrodiesel, but it is made of recently living biomass.

The terms renewable diesel and green diesel have been further distinguished based on the processing method to create the fuel with petrodiesel-like chemical composition.

"Renewable diesel" will refer to all diesel fuels derived from biomass that meet the standards of ASTM D975 and are not monoalkyl esters.

There are three primary methods for creating renewable diesel, hydrotreating, thermal conversion, and Biomass-to-Liquid. Renewable diesel can be made from the same feed stocks as biodiesel since both require the tricylglycerol containing material from biomass.

Renewable diesel blends follow the same nomenclature as biodiesel. Renewable diesel in its pure form is designated R100 while a blend comprised of 20% renewable diesel and 80% petrodiesel is called R20. Because renewable diesel is chemically the same as petrodiesel, it can be mixed with petrodiesel in any proportion but users may need to add an additive to address lubricity issue associated with compounds with no oxygen.

Renewable Diesel Production

Hydrotreating (hydroprocessing or hydrodeoxygenation) "Green Diesel" Process.

The hydrotreating process is a process utilized by petroleum refineries today to remove contaminants such as sulfur, nitrogen, condensed ring aromatics, or metals. In this process, feedstock is reacted with hydrogen under elevated temperature and pressure to change the chemical composition of the feedstock. In the case of renewable diesel, hydrogen is introduced to the feedstock in the presence of a catalyst to remove other atoms such as sulfur, oxygen and nitrogen to convert the triglyceride molecules into paraffinic hydrocarbons. In addition to creating a fuel that is very similar to petrodiesel, this process creates propane as a byproduct. Because this process requires fossil fuel-derived hydrogen, this process is not 100% renewable and this must be considered when calculating the energy return, greenhouse gas emissions (GHG) and carbon life cycle.

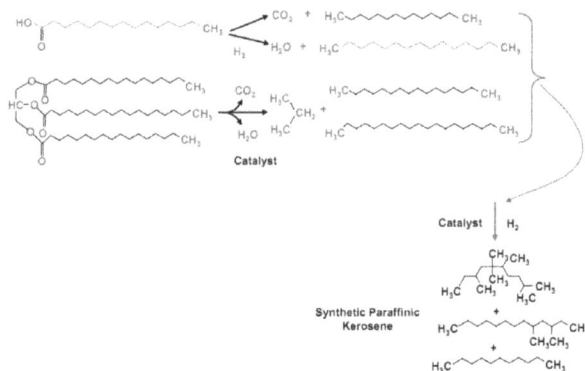

Hydrotreating Process

Many companies are utilizing this hydrotreating process as the basis for their renewable diesel projects. For instance, ConocoPhillips and Dynamic Fuels are working with Tyson Foods to convert waste animal fat into renewable diesel.

Since this process is currently used by many petroleum refineries, renewable diesel blends can be produced with existing refineries by co-processing the feedstock with petrodiesel. The advantages of this option when compared to constructing a stand alone operation are still under debate.

Thermal Depolymerization (hydrothermal processing, thermal conversion, cracking, pyrolysis, rapid thermal processing) "Renewable Diesel" Process.

Thermal depolymerization is another process that can convert biomass or other carbon-containing material into a "bio-oil" that is then refined into a petrodiesel-like fue. Conversion temperatures are typically 570-660 degrees Fahrenheit with a pressure range of 100-170 atmospheres. The process converts the large polymers (cellulose, hemi-cellulose, lignin, and proteins) of biomass into smaller molecules. As a result of this process, organic vapors, pyrolysis gases, and charcoal are produced. The vapors are condensed to produce pyrolysis oil or bio-oil.

Changing World Technologies are currently utilizing this method to process slaughterhouse waste and other carboncontaining solid waste to create a fuel that can meet the standards of both ASTM D396 and ASTM D975.

Biomass-to-Liquid (BTL) and Fischer-Tropsche

Yet another process for making renewable diesel fuel is to convert biomass (predominately cellulosic material) through high-temperature gasification into synthetic gas or "syngas," a gaseous mixture rich in hydrogen and carbon monoxide. The Fischer-Tropsche process is then used to catalytically convert the syngas to liquid fuel. This technology has been applied to coal (coal-to-liquids fuel or CTL), and natural gas (gas-to-liquids fuel or GTL) in addition to BTL.

When an organic material is burned, it can be completely oxidized or gasified to carbon dioxide and water, or it can be partially oxidized to carbon monoxide and hydrogen. The partially oxidized gasification reaction is accomplished by restricting the amount of oxygen during the combustion process. The resulting carbon monoxide and hydrogen mixture is the syngas which is the starting material for the Fischer-Tropsche process. The Fischer-Tropsche process is a set of chemical reactions that converts the syngas to into liquid hydrocarbons. The overall process can be seen below.

1) **Synthesis Gas Formation**

$$CH_n + O_2 \xrightarrow{\text{(Catalyst)}} \tfrac{1}{2}\, n\, H_2 + CO$$

2) **Fischer-Tropsch Reaction**

$$2n\, H_2 + CO \xrightarrow{\text{Catalyst}} -(CH_2-)_n- + H_2O$$

3) **Refining**

$$-(CH_2-)_n- \xrightarrow{\text{(Catalyst)}} \text{Fuels, lubricants, etc.}$$

BTL & Fischer Tropsche Process

Renewable diesel produced from BTL can be created using any source of biomass while other processes are limited to mainly lipids, oils obtained from recently living biomass.

BTL technology is mainly still in the research and development stages. Currently, Choren in Germany is working with Shell and VW on a BTL fuel it calls SunDiesel. Choren has built a plant in Freiberg, Germany with an estimated annual capacity of 18 million liters of SunDiesel.

Bioethers

Ethers are a class of carbon compounds that contain an ether group, which looks as follows.

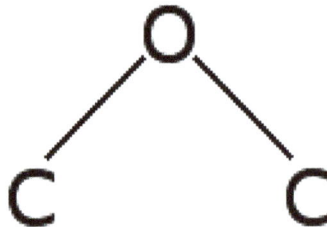

Either carbon in the basic ether above may be attached to other carbons as well. So, this image represents the most basic ether. All that is needed to make an ether is an oxygen atom with a hydrocarbon chain on either side.

Uses

Ethers are currently used as substitutes for lead to improve engine performance. They also decrease oxygen wear and toxic emissions, particularly ozone. The most commonly used fuel additive ethers are methyl-tertiary-butyl-ether (MBTE) and ethyl-tertiary-butyl-ether (ETBE).

Bioether

Bioether is made from either wheat or sugar beet. It can also be produced from the waste glycerol that results from the production of biodiesel. Bioether is likely to replace petro-ether as an additive to current fossil fuels. In the future, it is unlikely to be a fuel in and of itself because it has a low energy density. In fact, its energy density is about half that of standard diesel, which means you would need twice the volume of ether to go the same distance. Mostly likely, it will become an additive to other biofuels to reduce emissions.

	Dimethyl Ether	Diesel
Energy Density (MJ/kg)	28.8	42.7
Cetane Number	55-60	40-55

Despite its drawbacks, there is some interest in using ether, particularly dimethyl ether, as a replacement for diesel fuel. This is made possible by the fact that ether works well in compression ignition engines. The interest stems from the reduced particulate emissions and reduced pollution of ether. In urban environments, where refueling frequently is possible and pollution levels run high, the tradeoff of reduced emissions for more frequent refueling is an attractive one. The only major problem to ether in these settings is that it must be compressed to be a liquid. At standard temperatures and pressures, ether is a gas. To make it transportable, it must be compressed. Of course, compression has benefits as well. If the gas can be compressed enough, then more can be carried and refueling intervals will be longer. China is currently the leader in research regarding replacing diesel fuel with ether.

Biogas

The organic fraction of almost any form of biomass, including sewage sludge, animal wastes and industrial effluents, can be broken down through anaerobic digestion into methane and carbon dioxide mixture called as "biogas". Biogas is an environment friendly, clean, cheap and versatile fuel. Biogas is a valuable fuel which is produced in digesters filled with the feedstock like dung or sewage. The digestion is allowed to continue for a period of from ten days to a few weeks.

A methane digester system, commonly referred to as an anaerobic digester (AD) is a device that promotes the decomposition of manure or digestion of the organics in manure to simple organics and gaseous biogas products. There are three types of continuous digesters: vertical tank systems, horizontal tank or plug-flow systems, and multiple tank systems. Proper design, operation, and maintenance of continuous digesters produce a steady and predictable supply of usable biogas.

Several types of bio-digesters have been developed including the floating drum, the fixed dome, the bag, the plastic tube, the plug flow and the up-flow anaerobic sludge blanket digesters. Fig. below shows an on-farm biogas system.

Anaerobic decomposition is a complex process. Methane is produced in environments where organic matter accumulates, and oxygen is absent. The process by which anaerobic bacteria decompose organic matter into methane, carbon dioxide, and a nutrient-rich sludge involves a step-wise series of reactions requiring the cooperative action of several organisms. It occurs in three basic stages as the result of the activity of a variety of microorganisms. Initially, a group of microorganisms converts organic material to a form that a second group of organisms utilizes to form organic acids. Methane-generating (methanogenic) anaerobic bacteria utilize these acids and complete the decomposition process. In the first stage, a variety of primary producers (acidogens) break down the raw wastes into simpler fatty acids. In the second stage, a different group of organisms (methanogens) consume the organic acids produced by the acidogens, generating biogas as a metabolic byproduct. On average, acidogens grow much more quickly than methanogens. Finally, the organic acids are converted to biogas.

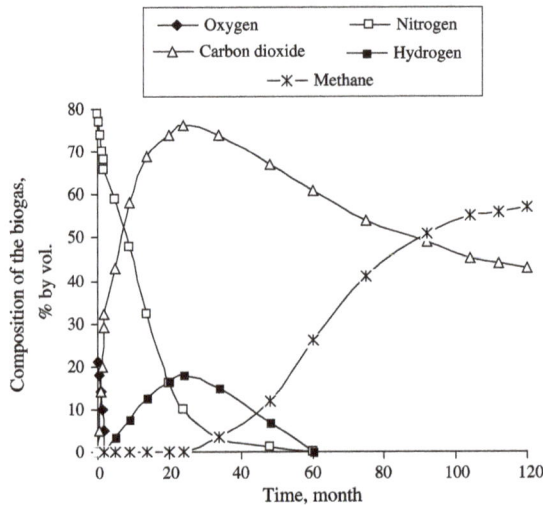

Production of biogas components with time in landfill

A variety of factors affect the rate of digestion and biogas production. The most important is temperature. Anaerobic bacteria communities can endure temperatures ranging from below freezing to above 330.4 K, but they thrive best at temperatures of about 309.9 K (mesophilic) and 327.6 K (thermophilic). Bacteria activity, and thus biogas production, falls off significantly between about 312.6 K and 324.9 K and gradually from 308.2 K to 273.2 K. Average 68% of the cultivated land produces grains with wheat ranking first, barley second, and corn third in developing countries. Agricultural solid residues are potential renewable energy resources. Wheat straw wastes represent a potential energy resource if they can be properly and biologically converted to methane. They are renewable and their net CO_2 contribution to the atmosphere is zero. Manures and manure/straw mixtures have been extensively investigated as sources of biogas.

In a process of manure and straw mixture digestion, for first 3 days, methane yield was almost 0% and carbon dioxide generation was almost 100%. In this period, digestion occurred as aerobic fermentation to carbon dioxide. The yields of methane and carbon dioxide gases were fifty–fifty at 11th day. At the end of the 20th day, the digestion reached the stationary phase. The methane content of the biogas was in the range of 73–79% for the runs, the remainder being principally carbon dioxide. During a 30-day digestion period, 80–85% of the biogas was produced in the first 15–18 days. This implies that the digester retention time can be designed to 15–18 days instead of 30 days.

Decomposition in landfills occurs in a series of stages, each of which is characterized by the increase or decrease of specific bacterial populations and the formation and utilization of certain metabolic products. The first stage of decomposition, which usually lasts less than a week, is characterized by the removal of oxygen from the waste by aerobic bacteria. In the second stage, which has been termed the anaerobic acid stage, a diverse population of hydrolytic and fermentative bacteria hydrolyzes polymers, such as cellulose, hemicellulose, proteins, and lipids, into soluble sugars, amino acids, long-chain carboxylic acids, and glycerol. Fig. above shows the behavior of biogas production with time, in terms of the biogas components. Fig. above indicates that the economic exploitation of CH_4 is worthwhile after one year from the start of the landfill operation. The main components of landfill gas are byproducts of the decomposition of organic material, usually in the form of domestic waste, by the action of naturally occurring bacteria under anaerobic conditions.

Agricultural residues are difficult to degrade bio-chemically. Pretreatment of straw by mechanical size reduction, heat treatment and/or chemical treatment with strong acids or bases usually improves the digestibility. Chemical pretreatment methods are bicarbonate treatment, alkaline peroxide treatment and ammonia treatment. Ammonia treatment has several advantages over the other treatments, such as being a source of nitrogen for biodegradation and the fact that no separate wastewater streams are generated from the pretreatment process.

Agricultural residues contain low nitrogen and have carbon-tonitrogen ratios (C/N) of around 60–90. The proper C/N ratio for anaerobic digestion is 25–35 ; therefore, nitrogen needs to be supplemented to enhance the anaerobic digestion of agricultural solid residues. Nitrogen can be added in inorganic form such as ammonia or in organic form such as livestock manure, urea, or food wastes. Once nitrogen is released from the organic matter, it becomes ammonium which is water soluble. Recycling nitrogen in the digested liquid reduces the amount of nitrogen needed.

Anaerobic biological treatment of the agricultural solid waste is a process which has received increased attention during the last few years. Conversion of these wastes to methane provides some energy and can have a beneficial effect on the environment, and during the digestion process bacteria in the manure are killed, which is a great benefit to environmental health. The production of methane during the anaerobic digestion of biologically degradable organic matter depends on the amount and kind of material added to the system.

Syngas

Syngas' or 'synthesis gas' is a combination of hydrogen, carbon monoxide, small quantities of carbon dioxide and other trace gases. Normally derived from feedstocks, Syngas contains carbon, such as biomass, natural gas, heavy oil and coal. In creates synthetic natural gas and producing methanol or ammonia.

Syngas is produced as a result of gasification of a carbon-containing fuel to a gaseous product that has heating value. If syngas contains nitrogen, it must be separated, as both nitrogen and carbon monoxide have similar boiling points and it will be difficult to recover pure carbon monoxide through cryogenic processing.

Syngas has 50% of the energy density of natural gas and hence it can be burnt and used as a fuel source. Refinement of syngas before use allows CO_2 to be stripped from the raw gas thereby enabling the use of CO_2 in enhanced oil recovery processes.

Production of Syngas

The production of syngas includes the following phases.

The Heating Phase

The first step is gasification, a thermo-chemical process in which carbon-rich feedstocks like petro-coke, biomass or coal are converted into a gaseous compound consisting of carbon monoxide and hydrogen under high-heat, high pressure, oxygen depleted conditions.

Very high temperatures of gasification, normally between 800 and 1500°C (1472 and 2732°F) are achieved with the help of an external heat source or through partial oxidation of feedstock which releases heat.

The Reaction Phase

The feedstock reacts with carbon dioxide, water vapor and oxygen during gasification. The reaction is triggered by thermal decomposition for oxygen-rich materials.

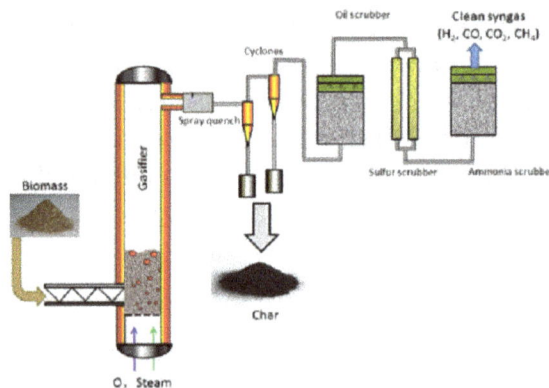

The process flow for Syngas production by gasification of biomass

The Purification Phase

The gas obtained from gasification is raw and not clean enough to use. A purification process is carried out to eliminate impurities like ash, tar, sulfur compounds, methane, water vapor and carbon dioxide. The hydrogen-oxygen proportion is adjusted after purification depending on the applications of the synthesis processes.

The Catalytic Phase

Metals such as iron, manganese, cobalt, copper and new complex molecules are formed when the syngas is in contact with different catalysts. Scientists are experimenting with several catalysts to find new ways of creating already existing molecular combinations. In this way, it is possible to create eco-friendly fuels from syngas.

Syngas Cleanup and Conditioning

Raw syngas obtained from the gasification process needs to be cleaned to eliminate the contaminants such as mercury, chlorides, ammonia, sulfur, fine particulates and other trace heavy metals to protect downstream processes and to meet environmental emission regulations.

Syngas may be conditioned to adjust the hydrogen-to-carbon monoxide ratio based on the downstream process application.

Typical syngas cleanup and conditioning processes include the following:

- Removing bulk particulates using cyclone and filters
- Wet scrubbing for eliminating chlorides, ammonia and fine particulates
- Removing trace heavy metal and mercury using solid absorbents
- Water gas shift for adjusting hydrogen-to-carbon monoxide ratio
- Catalytic hydrolysis for converting carbonyl sulfide to hydrogen sulfide
- Acid gas removal for extracting sulfur-bearing gases and carbon dioxide.

SynGas Fermentation

Syngas fermentation is a microbial process in which syngas is used as a carbon and energy source, and then converted into chemicals and fuels with the help of microorganisms. Methane, butyric acid, acetic acid, butanol and ethanol are the main products of syngas fermentation.

Acetogens such as Clostridium carboxidivorans, Eurobacterium limosum, Butyribacterium methylotrophicum and Peptostreptococcus products are involved in the production of chemicals and fuels.

Some of the key benefits of syngas fermentation process include the following:

- High reaction specificity

- Low temperature and pressure
- Does not require a specific ratio of CO to H_2
- Tolerate compounds having high sulfur content.

However, syngas fermentation has certain limitations such as inhibition of organisms, low volumetric productivity and gas-liquid mass transfer limitation.

Applications

Syngas can be used to produce a wide range of fertilizers, fuels, solvent and synthetic materials. Few examples are as follows:

- Steam for use in turbine drivers for electricity generation
- Nitrogen for use as pressurizing agents and fertilizers
- Hydrogen for electricity generation, use in refinery industry to extract more diesel and gasoline from crude oil and for a large variety of hydrogenation reactions where hydrogen is added to unsaturated hydrocarbons
- Ammonia for use as fertilizers and for the production of plastics like polyurethane and nylon.
- Methanol for the production of plastics, resins, pharmaceuticals, adhesives, paints and also as a component of fuels.
- Carbon monoxide for use in chemical industry feedstock and fuels
- Sulfur for use as elemental sulfur for chemical industry
- Minerals and solids for use as slag for roadbeds.

Solid Biofuels

Solid biofuels are produced from biomass through processing steps such as chopping, drying, chipping, grinding and densifying (pelletizing or briquetting). This improves the physical and chemical properties: i.e. particle size, moisture content and energy content. Densification is usually required for efficient and economical long distance transportation, bulk handling and storage. Additional processes are under development (torrefaction, steam processing) to further improve fuel quality by increasing energy content, mechanical durability and reducing water absorption. Solid biofuels are available in various forms including:

- Firewood: cut and/or split logs, preferably dried and usually with uniform length.

- Wood Chips: chipped wood with a defined size, a typical length of 5 to 50 mm, and a low thickness compared to length; produced by mechanical processing with sharp tools.

- Briquettes: densified (compressed) biomass fuel in a cubic or cylindrical form with a diameter of more than 25 mm.

- Wood Pellets: densified biomass fuel in a cylindrical form with a diameter up to 25 mm, typically 6 mm or 8mm, and length of 5 to 40 mm.

Origin of Solid Biomass Fuels

Sources of raw materials for the production of solid fuels are:

- Sawmill residues accrue from wood processing from local sawmills. The bark and deciduous/ coniferous proportion is low, which allows for a use of premium wood chips. Furthermore, sawmill by-products (especially saw dust and shavings) are used as raw material for the production of pellets.

- Roundwood as part of a tree trunk, which is not utilized for material use, due to a small diameter. This biomass source has a relatively low proportion of bark and needle.

- Wood residues from forest management are residues from various activities of forest management (young stand tending, thinning or felling), which cannot be used as furniture wood or industrial timber, due to small diameters, for example the top of the trees and tree branches. Wood residues normally have a high proportion of bark and needles / leafs. The use for industrial chips is therefore the standard variant, only some batches are also available for premium goods.

Raw material	Average Water content	Origin	Used for:
Saw mill residues	15 – 50%	Regional saw mills	Premium wood chips production, pellets
Round wood	20 – 50%	Forest, regional saw mills	Premium wood chips production
Forest residues	45 – 55% 30 – 40% if stored on forest street over summer	Private and municipal and federal forests	Industrial wood chips, and maybe premium wood chips
Landscaping material	45 – 60%	Private and municipal landscaping companies	Industrial wood chips
Short rotation coppices	45 – 55%	Short rotation coppices	Industrial wood chips, and premium wood chips
Stalk material	15 – 20%	Agricultural by-products	Straw fired plant

- Landscaping material / wood (LPM) arising during measures in the field of andscape and nature protection as well as of roadside vegetation maintenance. The woody residues usually have a high proportion of bark and needles or leaves. The biofuel is also often contaminated with soil or heavy metals, resulting from vehicle exhaust gases.

- Wood from Short Rotation Coppices (SRC) (mostly poplars or willows) can be used. Depending on the growth period and the rotation period, this can also be used for premium wood chips (benchmark: at least 5 years or a trunk diameter of > 10cm). Wood from SRC is also an alternative for pellet production.

- Agricultural by-products such as straw (stalks and leaves of threshed grains, legumes, oil and fibre plants) can make an important contribution to a sustainable energy supply.

Sustainability Aspects

The market for solid biomass has been growing rapidly in recent years. A sustainable development is not limited to ecological considerations. Besides the environmental protection aspect, the job creation and world nutrition ones are important too.

Thus the strength of the energetic use of biomass regarding the regional added value should be an integral part of the communication strategy. There is some controversy over the sustainability of bioenergy. To uphold the demand for solid biomass fuels, the concerns of the target groups should be addressed. Therefore, biomass suppliers should provide proof that the raw materials used meet the sustainability standards.

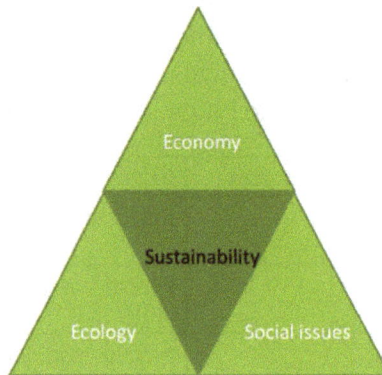

In order to continue to develop the biomass heating sector in a sustainable way, the following recommendations can be made:

- Utilise biomass (wood) from sustainably managed forests and ensure that new trees are planted in the place where forests have been harvested for bioenergy (SFM)

- Focus on the utilisation of wood left over from forestry or agricultural operations (thinnings, straw etc.) or industrial operations (industrial wood-chips, sawdust, etc.)

- Obtain proof of sustainable forest management through procurement from certified forests (FSC or PEFC). This will also stimulate the sector to certify the sustainability of the forest management further.

- Favour local resources; regionally available solid biofuels are the best choice as they:

  o Reduce transport distances and transport related GHG-emissions

  o Produce the highest possible local value added as they :

    ◇ Keep the money paid for energy services within the region instead of going abroad for fossil fuels

    ◇ Enable and secure local labour security and local infrastructure investments

    ◇ Help to enhance the stock value of local forests by financing proper forestry management.

- Provide best conditions to build up sustainable, resilient and distributed energy systems

- Enhance local acceptance of green energy investments and of subsidies for them

- Ensure a high energy conversion efficiency to minimize resource demand and emissions.

Wood Fuel

Wood constitutes the majority of biomass that is burned for fuel and comes in the forms firewood, charcoal, chips, pellets, and sawdust. The use of wood as a fuel for cooking, heating, and other applications dates back to well before humans when Neanderthals were the predominant species of hominid. In fact, the most troubling aspect of using wood as a fuel is generating the spark to start the fire. Otherwise, wood is readily available, abundant, and can even be collected from the ground if cutting tools are not available. Today, wood is even used in some electric generating applications.

Wood is reasonably energy dense. Hardwoods have an energy density of around 14-15 MJ/kg if burned with 100% efficiency. As with any fuel, however, the efficiency tends to be lower. Wood is actually more efficient than many fuels, with about 70% of the energy content (10 MJ/kg) recoverable on average.

The downside to wood is pollution. Not only does it produce more carbon dioxide than fuels like methane, it also produces other pollutants like soot, smoke, and PAHs. Research has lead to stoves that burn at extremely high temperatures (over 600 degrees Celsius). Temperatures this high actually allow the smoke itself to burn, which reduces emissions.

Wood fuel is often ignored in discussion about alternative energy. This is partly related to the fact that wood fuel was never really replaced by fossil fuels and partly due to misconception that it harms the environment by emitting carbon dioxide and contributing to deforestation. But if proper technology is used and if wood fuel is sourced from sustainably managed forests, it is a carbon neutral source of energy and does not harm the environment in terms of destroying wildlife habitat.

Wood fuel is available as firewood, pellets, chips, charcoal and sawdust. It is mainly used for space and water heating. Wood burning to power steam engines to generate electricity is rare.

Advantages

- Availability. Wood fuel is easily available either in the form of firewood or wood wastes.

- Low cost. Wood fuel is inexpensive, especially in comparison to fossil fuels. And unlike other alternatives to fossil fuels, it does not require high initial investment.

- Environmentally friendly. As mentioned earlier, wood fuel is not harming the environment if it is sourced in sustainably managed forests and if the right technology is used, for example efficient wood burning stoves. In fact, the same trees that are harvested for firewood and teak outdoor furniture provide a habitat to wildlife and reduce carbon dioxide in the atmosphere while growing. Wood burning does emit carbon dioxide but since the amounts of carbon emitted equal to the amounts of carbon absorbed while the tree was growing, the total carbon emissions equal zero.

- Sustainability. Wood fuel is a sustainable source of energy because each tree that is cut down can be replaced.

Disadvantages

- Inconvenience. In comparison to other forms of heating, wood fuel is often less convenient because most stoves and furnaces require refilling every few hours. Automatised systems are available as well but they are quite expensive.

- Environmental concerns. Deforestation and in the first place, illegal logging is a serious environmental concern. In the developed countries, the rate of illegal logging is relatively low but it remains high in the developing nations. Unsustainable wood harvesting does not only contribute to the rising levels of carbon dioxide in the atmosphere but it also threatens the local ecosystems and biodiversity.

- Low benefits in terms of lowering carbon emissions. Trees absorb huge amounts of carbon dioxide and other greenhouse gases while growing but all absorbed gases are released back in the atmosphere during wood burning.

Bagasse

Bagasse or megass is a fibre remaining after the extraction of the sugar-bearing juice from sugarcane.

The word bagasse, from the French bagage via the Spanish bagazo, originally meant "rubbish," "refuse," or "trash." Applied first to the debris from the pressing of olives, palm nuts, and grapes, the word was subsequently used to mean residues from other processed plant materials such as sisal, sugarcane, and sugar beets. In modern use, the word is limited to the by-product of the sugarcane mill.

Bagasse is burned as fuel in the sugarcane mill or used as a source of cellulose for manufacturing animal feeds. Paper is produced from bagasse in several Latin American countries, in the Middle East, and in sugar-producing countries that are deficient in forest resources. Bagasse is the essential ingredient for the production of pressed building board, acoustical tile, and other construction materials and can be made into a number of biodegradable plastics. Bagasse is also employed in the production of furfural, a clear colourless liquid used in the synthesis of chemical products such as nylons, solvents, and even medicines. Bagasse is readily available as a waste product with a high sugar content and has potential as an environmentally friendly alternative to corn as a source of the biofuel ethanol (ethyl alcohol).

Uses of Bagasse

This pulp has long been burned by sugarcane processors as fuel to run factories – in fact, one of the largest biomass energy facility (a power plant that burns biological waste for energy) in the U.S. is run by a sugar company. Bagasse is much cleaner burning than fossil fuels and is considered carbon neutral since the amount of carbon released is the same as the amount of carbon the plants absorbed during growth. There is also research being done on using bagasse as a source of ethanol with the goal of reducing greenhouse gas emissions by as much as 85% over gasoline.

Another use for bagasse, and the most important one for everyday consumers, is as a source of pulp for paper. The fibers in bagasse can be used as an alternative to wood in the production of paper products like tissues, napkins, toilet paper, stationary, and cardboard without any loss in the quality of the product. It's also sturdy enough to replace styrofoam and plastic for use in disposable plates, cups, and takeout containers.

Importance of Bagasse

From seed to final product, bagasse is an environmentally friendly alternative to everyday items made from plastic and wood pulp. Because it's a byproduct of sugar production, it does not require a heavy investment of new resources to produce, and bagasse products are actually a way to make use of material that would otherwise be discarded. Sugarcane is also much faster growing than trees, which means that the use of bagasse for the manufacturing of paper makes it easy to reduce our reliance on harvesting wood.

One of the reasons bagasse is so important is that these products can replace the less environmentally-friendly versions we use every day. Many restaurants have replaced plastic or styrofoam take-out containers with bagasse, and you can also find disposable kitchenware like plates, napkins, bowls, and even toilet paper all made from bagasse. And unlike other kinds of paper, bagasse kitchen products don't need a plastic lining to stay waterproof, which means they can easily be composted, and these cups and plates will hold food and liquids up to 200 °F.

Disposal of Bagasse

Because bagasse is plant-based, it is easily compostable. Bagasse napkins, cups, or packaging will decompose in an average of 2 to 4 months in a home compost bin and in even less time in an industrial compost site. Just remember if you have bagasse products to dispose of to make sure they find their way into the compost: landfills cover waste so it doesn't have access to the air, moisture, and microorganisms necessary for decomposition, meaning bagasse products you throw in the trash will take a very long time to break down. If disposed of properly, though, bagasse is a smart, eco-friendly alternative to plastic.

Cow Dung

Cow dung, manure, or feces is indigestible plant material released on to the ground from the intestine of a cow. Feces is generally not a favourite topic of conversation, whether it comes from an animal or a human. Cow dung is worth discussing, though. It's a useful material and helps us in a variety of ways. It's also a plentiful and renewable resource. It's a shame when it's wasted.

Cow manure has a soft texture and tends to be deposited in a circular shape, which gives dung patches their alternate names of cow pies and cow pats. The manure is used as a rich fertilizer, an efficient fuel and biogas producer, a useful building material, a raw material for paper making, an insect repellent, and a disinfectant. Cow dung "chips" are used in throwing contests and cow pie bingo is played as a game. The manure also plays an essential role in the lives of various animals, plants, and microbes, including dung beetles and the Pilobolus fungus.

Cow dung drying in stacks for fuel

Fuel and Biogas from Cow Manure

Dried cow dung is an excellent fuel. In some cultures dung from domestic cows or buffalo is routinely collected and dried for fuel, sometimes after being mixed with straw. Pieces of dung are lit to provide heat and a flame for cooking. Dried dung has lost its objectionable odour.

Production and uses of a Biogas

The general process for making an anaerobic digester for cow dung starts with placing dung and water in an airtight container. The container must be kept warm and left undisturbed so that bacteria can do their work. The gas that is produced is withdrawn through a tube and stored.

A Highland cow calf

Once a biogas has formed, it can be reacted with oxygen to produce energy. The gas can be used to cook food, heat water in a boiler, and replace conventional fuel in motor vehicles. In addition, the energy in a biogas can be used to produce electricity.

Biogas produced from cow dung generally consists of methane, carbon dioxide, and other components, such as hydrogen sulphide. Since there is so much methane in the gas, it's important that it doesn't escape into the environment. Methane is a major greenhouse gas and contributes to global warming.

References

- Aviation-fuel, homebuilt-aircraft: experimentalaircraft.info, Retrieved 29 March 2018

- The-differences-between-kerosene-jet-fuel-12003828: hunker.com, Retrieved 28 May 2018

- Bagasse, technology: britannica.com, Retrieved 21 June 2018

- What-is-bagasse: greenhome.com, Retrieved 11 April 2018

- The-Many-Uses-of-Cow-Dung, agriculture: owlcation.com, Retrieved 28 May 2018

Chapter 3

Biomass and its Types

Biomass refers to the generation of energy by burning wood and other organic matter. Biomass is directly used via combustion to generate heat or converted into various forms of biofuel. The topics elaborated in this chapter on woody and non-woody biomass will help in providing a better perspective of biomass and its various types.

The term "biomass" refers to organic matter that has stored energy through the process of photosynthesis. It exists in one form as plants and may be transferred through the food chain to animals' bodies and their wastes, all of which can be converted for everyday human use through processes such as combustion, which releases the carbon dioxide stored in the plant material. Many of the biomass fuels used today come in the form of wood products, dried vegetation, crop residues, and aquatic plants. Biomass has become one of the most commonly used renewable sources of energy in the last two decades, second only to hydropower in the generation of electricity. It is such a widely utilized source of energy, probably due to its low cost and indigenous nature, that it accounts for almost 15% of the world's total energy supply and as much as 35% in developing countries, mostly for cooking and heating.

Biomass is one of the most plentiful and well-utilised sources of renewable energy in the world. Broadly speaking, it is organic material produced by the photosynthesis of light. The chemical material (organic compounds of carbons) are stored and can then be used to generate energy. The most common biomass used for energy is wood from trees. Wood has been used by humans for producing energy for heating and cooking for a very long time.

Biomass has been converted by partial-pyrolisis to charcoal for thousands of years. Charcoal, in turn has been used for forging metals and for light industry for millenia. Both wood and charcoal formed part of the backbone of the early Industrial Revolution (much northern England, Scotland and Ireland were deforested to produce charcoal) prior to the discovery of coal for energy.

Wood is still used extensively for energy in both household situations, and in industry, particularly in the timber, paper and pulp and other forestry-related industries. Woody biomass accounts for over 10% of the primary energy consumed in Austria, and it accounts for much more of the primary energy consumed in most of the developing world, primarily for cooking and space heating.

It is used to raise steam, which, in turn, is used as a by-product to generate electricity Considerable research and development work is currently underway to develop smaller gasifiers that would produce electricity on a small-scale. For the moment, however, biomass is used for off-grid electricity generation, but almost exclusively on a large-, industrial-scale.

There are two issues that affect the evaluation of biomass as a viable solution to our energy problem: the effects of the farming and production of biomass and the effects of the factory conversion of biomass into usable energy or electricity. There are as many environmental and economic benefits as there are detriments to each issue, which presents a difficult challenge in evaluating the potential success of biomass as an alternative fuel. For instance, the replacement of coal by biomass could result in "a considerable reduction in net carbon dioxide emissions that contribute to the greenhouse effect." On the other hand, the use of wood and other plant material for fuel may mean deforestation. We are all aware of the problems associated with denuding forests, and widespread clear cutting can lead to groundwater contamination and irreversible erosion patterns that could literally change the structure of the world ecology.

Biomass Power

Biomass power is carbon neutral electricity generated from renewable organic waste that would otherwise be dumped in landfills, openly burned, or left as fodder for forest fires.

When burned, the energy in biomass is released as heat. If you have a fireplace, you already are participating in the use of biomass as the wood you burn in it is a biomass fuel.

In biomass power plants, wood waste or other waste is burned to produce steam that runs a turbine to make electricity, or that provides heat to industries and homes. Fortunately, new technologies — including pollution controls and combustion engineering — have advanced to the point that any emissions from burning biomass in industrial facilities are generally less than emissions produced when using fossil fuels (coal, natural gas, oil). ReEnergy has included these technologies in our facilities.

Technologies For Bio-power

Biomass power can be generated by using two technologies one is thermo-chemical and other is biochemical. In thermo-chemical: direct firing, co-firing, gasification and pyrolysis techniques can be used for woody biomass, while in bio-chemical anaerobic digestion and fermentation techniques can be used for non woody biomass.

Direct Combustion

Direct combustion is similar to thermal power generation, in which biomass is burnt in the boiler and produce steam to run the turbine to produce electricity. It is based on rankine steam cycle. The chemical reaction is:

$$C_xH_yO+O_2 \rightarrow CO_2+H_2O+Heat$$

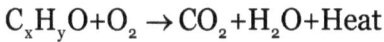

The capacity range of these plants is ranges from 0.5 MW to 10 MW and the efficiency range of these plants are 15-25%.

This technology disposes of large amount of residue and wastes. The block diagram of direct firing is shown in figure below.

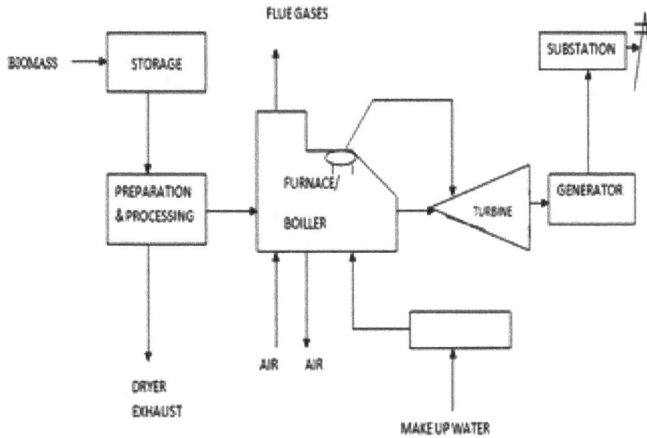

Block diagram of direct firing

Co-firing

Co-firing is mixing of a percentage of biomass with the coal in coal fired stations. Co-firing can also be used to improve the combustion of fuels with low energy content. The different options for co-firing:

- Biomass blending with coal or Direct firing – It is simple and least cost approach. Biomass fuels are blended with coal and blend is sent to the firing system.

- Separate Injection or Indirect firing – In this approach the biomass is separately injected into the boiler without impacting the coal delivery process. This method involves more equipment than the first approach.

- Gasification based Co-firing – In this approach, biomass is first fed to gasifier to generate producer gas and then it is fired in boiler.

- Gasification

  Gasification is a technique in which thermo chemical conversion of solid biomass into highly combustible gas for burning is obtained by partial oxidation under high temperature. The gas obtained from the gasification process is a mixture of CO, H2 and CH4 with CO2 andN2. The gas can be used in internal combustion engine or in gas turbine.

- Pyrolysis

  Pyrolysis is biomass conversion technique, through which biomass is converted to liquid, solid and gaseous fractions by heating the biomass in the absence of air or oxygen. In this volatile component of the waste are vaporized by heating, leaving residue consist of char and ash.

- Anaerobic Digestion

  In anaerobic digestion, organic material is broken down by bacteria in the absence of oxygen, to produce methane rich biogas. The solid waste left can be used as compost in fields. There are four biological and chemical stages of anaerobic digestion shown in figure below.

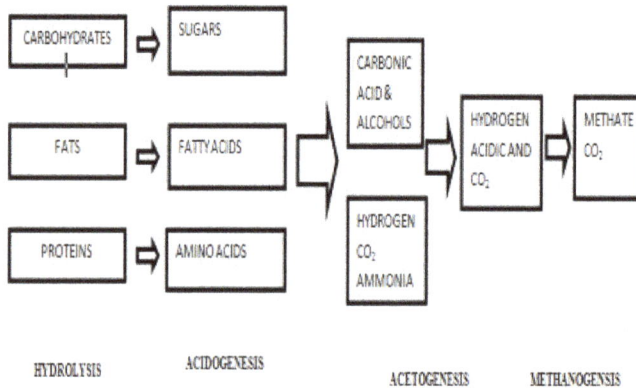

Four stages of anaerobic digestion

- Fermentation

  Fermentation is a technique in which a mixture of 60% methane and 40% CO_2 produced through anaerobic fermentation of material like plants vegetable waste and animal waste etc. This method requires large installation cost, longer reaction time, high amount of water and large area for installation of plant. Production of ethanol and methanol:

  o Ethanol – Ethanol is produced from the biomass like sugarcane, starches and cellulose i.e. wood and agricultural residues.

  o Methanol – High cellulose content materials such as wood and agricultural residues are suitable for methanol production.

Conversion of Biomass into Energy

Burning

This is a very common way of converting organic matter into energy. Burning stuff like wood, waste and other plant matter releases stored chemical energy in the form of heat, which can be used to turn shafts to produce electricity. Let's see this simple illustration of how biomass is used to generate electricity.

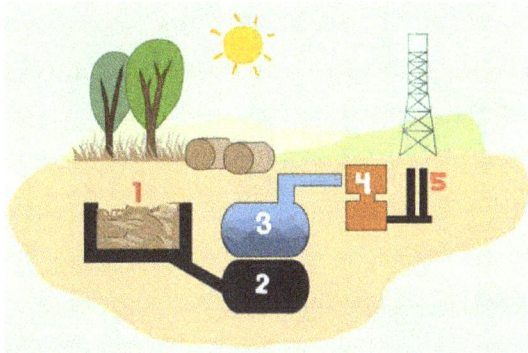

- Energy from the sun is transferred and stored in plants. When the plants are cut or die, wood chips, straw and other plant matter is delivered to the bunker.

- This is burned to heat water in a boiler to release heat energy (steam).

- The energy/power from the steam is directed to turbines with pipes.

- The steam turns a number of blades in the turbine and generators, which are made of coils and magnets.

- The charged magnetic fields produce electricity, which is sent to homes by cables.

Other ways in which organic matter can be converted into energy includes the following.

Decomposition

Things that can rot, like garbage, human and animal waste, dead animals and the like can be left to rot, releasing a gas called biogas (also known as methane gas or landfill gas). Methane can be captured by a machine called Microturbine and converted into electricity. Sometimes, animal waste (poop) can also be converted into methane by a machine called 'Anaerobic Digester'.

Fermentation

Ethanol can be produced from crops with lots of sugars, like corn and sugarcane. The process used to produce ethanol is called gasification.

Biomass Resources

All processing of biomass yields by-products and waste streams collectively called residues, which have significant energy potential. A wide range of biomass resources are available for transformation into energy in natural forests, rural areas and urban centres.

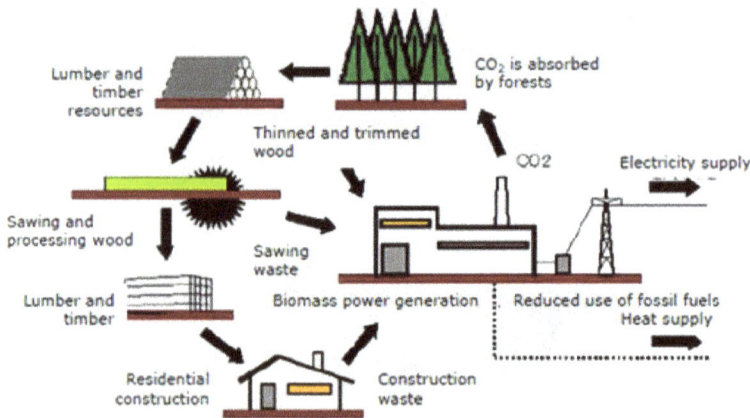

A host of natural and human activities contributes to the biomass feedstock

Pulp and Paper Industry Residues

The largest source of energy from wood is the waste product from the pulp and paper industry called black liquor. Logging and processing operations generate vast amounts of biomass residues. Wood processing produces sawdust and a collection of bark, branches and leaves/needles. A paper mill, which consumes vast amount of electricity, utilizes the pulp residues to create energy for in-house usage.

Forest Residues

Forest harvesting is a major source of biomass for energy. Harvesting may occur as thinning in young stands, or cutting in older stands for timber or pulp that also yields tops and branches usable for bioenergy. Harvesting operations usually remove only 25 to 50 percent of the volume, leaving the residues available as biomass for energy. Stands damaged by insects, disease or fire are additional sources of biomass. Forest residues normally have low density and fuel values that keep transport costs high, and so it is economical to reduce the biomass density in the forest itself.

Agricultural or Crop Residues

Agriculture crop residues include corn stover (stalks and leaves), wheat straw, rice straw, nut hulls etc. Corn stover is a major source for bioenergy applications due to the huge areas dedicated to corn cultivation worldwide.

Urban Wood Waste

Such waste consists of lawn and tree trimmings, whole tree trunks, wood pallets and any other construction and demolition wastes made from lumber. The rejected woody material can be collected after a construction or demolition project and turned into mulch, compost or used to fuel bioenergy plants.

Energy Crops

Dedicated energy crops are another source of woody biomass for energy. These crops are fast-growing plants, trees or other herbaceous biomass which are harvested specifically for energy production. Rapidly-growing, pest-tolerant, site and soil-specific crops have been identified by making use of bioengineering. For example, operational yield in the northern hemisphere is 10-15 tonnes/ha annually. A typical 20 MW steam cycle power station using energy crops would require a land area of around 8,000 ha to supply energy on rotation.

Herbaceous energy crops are harvested annually after taking two to three years to reach full productivity. These include grasses such as switchgrass, elephant grass, bamboo, sweet sorghum, wheatgrass etc.

Short rotation woody crops are fast growing hardwood trees harvested within five to eight years after planting. These include poplar, willow, silver maple, cottonwood, green ash, black walnut, sweetgum, and sycamore.

Industrial crops are grown to produce specific industrial chemicals or materials, e.g. kenaf and straws for fiber, and castor for ricinoleic acid. Agricultural crops include cornstarch and corn oil? soybean oil and meal? wheat starch, other vegetable oils etc. Aquatic resources such as algae, giant kelp, seaweed, and microflora also contribute to bioenergy feedstock.

Advantages of Biomass

1. It is a renewable source of energy:

A lot of organic and agricultural wastes are generated each day. Biomass can be produced from these wastes. As such, biomass is a resource that is readily available and virtually inexhaustible. Biomass energy generates power using renewable assets which come from sources such as wood waste, tree buildup, urban waste and handled wood pellets.

2. Its carbon content is part of the source:

Burning coal or gas releases carbon into the atmosphere. On the other hand, biomass sources already have carbon present in them. Plants, for instance, already have energy stored in them. They absorb energy from the sun through photosynthesis. The stored energy is then released as heat when biomass is burned.

Carbon dioxide is released when biomass is burned. Plants, however, also make use of carbon dioxide from the atmosphere in order to grow. The carbon dioxide they take is returned to the air when the plants are burned.

3. It lessens dependence on fossil fuels:

Biomass has the capability to replace all other sources of fuel. Since it relies on natural materials to generate power, there will be less reliance on energy created by assets that are non-renewable such as coal and gas. The use of renewable energy resources is good for the planet as well.

4. It is widely available:

The concept behind biomass is turning trash into energy. We produce lots of trash each day and how great would it be if we can harness this to produce something good? It's possible to supply power to at least 1.3 million homes in the US with the amount of trash we produce each day.

Organic waste is also a source of biomass. Where does waste such as dead trees and mowed grass go? Rather than send them to the landfills, why not harness them for energy?

5. It lessens the amount of waste we produce:

We produce a huge amount of waste each day. Some of the waste we produce are biodegradable, recyclable and are also toxic. Rather than let these sit on landfills, why don't we use them to produce energy? Many landfills are now suffering from too much trash so why don't we lessen that load by making use of trash that is still beneficial?

Disadvantages of Biomass

1. It leads to deforestation:

Wood is a major source for biomass energy. A major source for wood is, of course, trees. A large number of trees are being cut down in order to produce a sufficient amount of power. Continuing to operate on such a large scale will eventually result in the destruction of our forests.

2. It doesn't offer a 100% clean burn:

One of the highlights of biomass energy is that it's friendly for the environment. While mostly true, it doesn't erase the fact that burning wood and other organic matter does create pollution. Various compounds are also released when burning biomass sources. Yes, biomass has been touted as renewable but we have yet to prove that it absolutely does nothing wrong to the environment.

3. It needs a lot of space to produce:

Most of the biomass facilities are located in urban areas. As such, they contribute to more traffic in the area as well as more pollution. We are slowly becoming a very urbanized world and there are projections that suggest it will be the area where most of us would live by the end of the century. We are already feeling the disadvantages of urbanization and biomass facilities seemingly make it worse.

4. It is not cheap:

The energy plants that need to be built for biomass production are very expensive. The cost of gathering the required resources as well as the expenses needed for transportation are also quite high.

5. It contributes to pollution:

Biomass sources need to be transported. The use of vehicles has been linked with the increase in air pollution. Burning biomass releases carbon which is bad for the atmosphere. Yes, there are many benefits to the use of biomass and it has been hailed as better for the environment but it's contribution to pollution shouldn't be neglected.

Woody Biomass

Woody biomass is defined as the by-product of management, restoration, and hazardous fuel reduction treatments, as well as the product of natural disasters, including trees and woody plants (limbs, tops, needles, leaves, and other woody parts, grown in a forest, woodland, or rangeland environment).

The term woody biomass is perhaps best used to mean the material obtained from trees or the products of trees that has accumulated to a sufficient quantity that it is a hazard or disposal problem, or from trees specifically managed for biomass markets. This includes the materials listed below as sources of woody biomass, but it does not include waste paper or chemical residues from the pulp and paper industry that already have viable markets. The above definition also precludes wood and wood residues that would otherwise fit the definition except that they already are used to produce higher-value products, such as sawmill residues for particleboard and other composite panels.

Woody biomass utilization is sometimes narrowly defined to mean the use of wood as a source or feedstock solely for the production of energy (heat and electricity). This is short-sighted and often hinders the discovery of its full potential. It is important to remember that although woody biomass is low in value and quality it has potential as a feedstock for energy production as well as for higher value manufactured goods.

Sources of Woody Biomass

Woody biomass is the solid portion of stems and branches from trees or residue products made from trees. Woody biomass can come from a variety of sources, including:

| dense timber stand with high fuel hazard | sawmill residues | Dedicated poplar forest |

- Non-timber tree removal – removing dead and dying trees, unwanted urban trees, or trees impeding land development,

- Forest management harvesting – the removal of small diameter trees from overpopulated stands for wildfire hazard fuel reduction, pre-commercial thinning of timber stands, or forest health improvement,

- Timber harvesting and logging residues – non-merchantable wood including branches, undersized trees, and non commercial species removed during typical timber harvesting operations,

- Sawmill and other wood manufacturing residues – includes bark, undersized and defective wood pieces (e.g. lumber edging and end trimming), sawdust, and other wood waste,

- Landfill diversion – wood debris from tree removal and pruning, construction, demolition, discarded shipping materials (boxes, crates, pallets), and other trashed wood products,

- Chaparral management – removal of excess woody shrubs and plants for wildfire fuel hazard reduction or other vegetation management goals.

- Dedicated forests (plantations)—fast growing trees grown specifically for biomass markets.

Biomass Utilization

Woody biomass has the potential to be used to make any of the products that are already produced or manufactured from wood. And, just like wood it benefits from the same advantages. But because woody biomass is by definition a low quality raw material there are many challenges to using the resource effectively and efficiently. These advantages and some potential challenges are summarized below.

Advantages

- Is a non-food, organic material that as a feedstock will not compete with agricultural interests for growing food crops.

- Is a renewable resource.

- Has positive benefits to the environment because of the lower energy require-ments to manufacture products than comparable non-wood materials (e.g. wood vs. steel).

- Can reduce wildfire hazard when it is removed from the forest or wildland ur-ban interface.

- When burned to produce energy, emits an amount of carbon dioxide that is comparable to the amount of CO_2 released by wood during natural degrada-tion. Since trees take in CO_2 during photosynthesis, using wood to produce en-ergy is considered "carbon neutral."

Challenges

- Limited availability – difficult to obtain enough feedstock near large processing plants to take advantage of the "economies of scale."

- High cost – associated with the harvest, transportation and storage of woody biomass.

- Low-energy density – woody biomass has a much lower energy density (unit of energy per unit of volume) than fossil fuels.

- Technological difficulties – the recalcitrant nature of wood and its complex chemistry make it a technologically difficult feedstock to use in chemical and energy processing.

- Competition – the growth of woody biomass markets may divert higher-quality wood away from traditional wood product markets.

Woody Biomass for Commercial Forest Products

Woody biomass in tree form cannot be used to make lumber, if it could it would be clas-sified as timber and not woody biomass. By definition, woody biomass is unsuitable for lumber. Woody biomass in tree form comes from trees that are too small to make standard lumber sizes or trees that are too low in quality (e.g. decayed) to achieve minimum lumber quality standards. Woody biomass in non-tree form is found in a size and shape unsuitable for lumber. However, many other commercial forest products can be made from woody biomass. The range of products that could be produced from woody biomass includes:

- Soil amendments: woody biomass roughly milled into small particles can be added directly to soil as a carbon rich, degradable buffer or to other organic materials high in nitrogen content to form a balanced compost,

- Landscape materials: bark and wood chips can be processed into uniform sizes for decorative ground cover, mulch, or playground fill,

- Firewood: woody biomass in stem form can be processed into standard fire-wood sizes by cutting to length and splitting,

- Posts and poles: woody biomass in stem form can be sorted to obtain stems large enough and with enough sound wood that they can be cut to length and peeled to a uniform diameter of standard post or pole size,

- Wood fiber resource: woody biomass treated to a strict preparation process can be cleaned up and sized appropriately to be the feedstock for composite panel production (particleboard, oriented-strand board, and fiberboard) , wood-plastic lumber, or pulp for paper,

- Feedstock for bio-refineries: contaminant free woody biomass could be the feedstock to produce numerous organic chemicals, including biofuels.

The above list is ordered in ascending value of the product produced and also in increasing cost of feedstock preparation. The use of woody biomass as a wood fiber resource for composite products or as a feedstock for bio-refineries is unlikely at present because of the high processing costs associated with procurement, transportation and preparation of the feedstock. But as technological improvements lower processing costs and competing raw materials increase in cost, the time may come in the not near future where these higher value products become economical.

Woody Biomass for Bioenergy

Woody biomass and bioenergy are in some circles nearly synonymous terms. This is not surprising considering the heightened interest in alternative energy, the impacts of energy production and use on climate change, and the definite advantages related to expanding the use of woody biomass for energy. The technology exists to produce heat energy directly from woody biomass or to produce intermediate biofuels designed to be stored and transported long distances. Bioenergy categories suitable for woody biomass include the following.

- Heat Energy – Through the exothermic combustion process, wood or woody biomass, is converted into the primary products of carbon dioxide, water, inorganic ash, and various gaseous and particulate emissions while giving off about 8,000 BTU's of heat for every pound of dry wood burned.

- Electrical Energy – Coupling the combustion process with a steam boiler and using the produced steam to drive an electrical turbine is a well proven method of producing electricity from woody biomass.

- Biofuels – Many types of woody biomass derived biofuels are possible but of the following fuels only the first three are in common usage and have proved to be economically viable. The remaining fuels listed are the subject of much research and development interests.

o   Solid or milled wood – wood in any size or shape can be directly combusted to produce heat and as such is a biofuel firewood and wood chips are common market categories.

o   Densified wood – wood particles are compressed into a smaller volume of a specific size and shape (pellets, logs, bricks, etc.) to increase the fuel density (Btu's per unit volume), ease of transportation, enhance storage durability, or improve other burning characteristics.

o   Charcoal – Produced by subjecting wood to a slow pyrolysis process (heating at 700 -900°F in the absence of oxygen for many hours) that thermally degrades the wood into an aqueous liquid fraction (tar), a gaseous fraction, and a solid fraction consisting mostly of carbon (char) that is formed into charcoal.

o   Bio-oil – Produced from the liquid fraction of wood pyrolysis. Rapid pyrolysis at high temperatures, rapid heating rates and short residence times maximizes the yield of bio-oil and minimizes the quantity of char produced.

o   Alcohol – Produced by subjecting wood to a hydrolysis/fermentation process. In the hydrolysis step wood particles are broken down into aqueous solutions of simple sugars, usually using acids, enzymes, or both. During the fermentation step, yeast converts the simple sugars into alcohol. Ethanol is the alcohol most commonly produced but other alcohols are possible.

o   Producer gas – A combustible gas of carbon dioxide, hydrogen, and various hydrocarbons that is produced by subjecting wood to a gasification process (heating at about 1400°F with a controlled, limited amount of air or oxygen) that converts the wood into a gaseous fraction (producer gas), char, tar, and ash. The producer gas can be upgraded into syngas or many higher value chemicals through catalytic conversion.

o   Bio-diesel – catalytic conditioning of syngas that was derived from the gasification of woody biomass can be directed towards the production of synthetic bio-diesel with properties similar to bio-diesel produced by the transesterification of triglycerides (e.g. vegetable oils and fats) or diesel derived from fossil fuels.

o   Drop in fuels – the next generation of bio-based fuels that are completely interchangeable and compatible with conventional fuels.

Coconut

Coconuts are produced in 92 countries worldwide on about more than 10 million hectares. Indonesia, Philippines and India account for almost 75% of world coconut production with Indonesia being the world's largest coconut producer. A coconut plantation is analogous to energy crop plantations, however coconut plantations are a source of wide variety of products, in addition to energy. The current world production of coconuts has the potential to produce electricity, heat, fiberboards, organic fertilizer, animal feeds, fuel additives for cleaner emissions, health drinks, etc.

The coconut fruit yields 40 % coconut husks containing 30 % fiber, with dust making up the rest. The chemical composition of coconut husks consists of cellulose, lignin, pyroligneous acid, gas, charcoal, tar, tannin, and potassium. Coconut dust has high lignin and cellulose content. The materials contained in the casing of coco dusts and coconut fibers are resistant to bacteria and fungi.

Coconut husk and shells are an attractive biomass fuel and are also a good source of charcoal. The major advantage of using coconut biomass as a fuel is that coconut is a permanent crop and available round the year so there is constant whole year supply. Activated carbon manufactured from coconut shell is considered extremely effective for the removal of impurities in wastewater treatment processes.

Coconut Shell

Coconut shell is an agricultural waste and is available in plentiful quantities throughout tropical countries worldwide. In many countries, coconut shell is subjected to open burning which contributes significantly to CO_2 and methane emissions. Coconut shell is widely used for making charcoal. The traditional pit method of production has a charcoal yield of 25–30% of the dry weight of shells used. The charcoal produced by this method is of variable quality, and often contaminated with extraneous matter and soil. The smoke evolved from pit method is not only a nuisance but also a health hazard.

The coconut shell has a high calorific value of 20.8MJ/kg and can be used to produce steam, energy-rich gases, bio-oil, biochar etc. It is to be noted that coconut shell and coconut husk are solid fuels and have the peculiarities and problems inherent in this kind of fuel. Coconut shell is more suitable for pyrolysis process as it contain lower ash content, high volatile matter content and available at a cheap cost. The higher fixed carbon content leads to the production to a high-quality solid residue which can be used as activated carbon in wastewater treatment. Coconut shell can be easily collected in places where coconut meat is traditionally used in food processing.

Coconut Husk

Coconut husk has high amount of lignin and cellulose, and that is why it has a high calorific value of 18.62MJ/kg. The chemical composition of coconut husks consists of cellulose, lignin, pyroligneous acid, gas, charcoal, tar, tannin, and potassium. The predominant use of coconut husks is in direct combustion in order to make charcoal, otherwise husks are simply thrown away. Coconut husk can be transformed into a value-added fuel source which can replace wood and other traditional fuel sources. In terms of the availability and costs of coconut husks, they have good potential for use in power plants.

Oil Palm

The oil palm biomass can be categorised into two: solid and liquid biomass. The production of both solid and liquid biomass is the result of the process of extracting oil from oil palm. As a rich source of biomass, the oil palm consists of the following components: empty fruit bunches (EFB), palm fronds, tree trunks, mesocarp fibres (MF), palm kernel shells (PKS), and palm oil mill effluents (POME).

Different components of oil palm biomass have different potential uses, including as fuel for electricity generation, biochemical products, fertiliser, and biofuel. This also means that different components may require different processing methods to be converted into each of their usable forms.

Palm Fronds

For per hectare of oil palm plantation, about 10 tonnes of dry palm fronds are produced. The palm frond consists of two main parts: petiole and leaflets. Palm fronds are cut during harvesting of fruit bunches. In Malaysia, apart from being used as fertilisers, fronds from oil palm trees are also processed into fibres, which in turn are used to make animal feeds for livestock. Additionally, palm fronds are used as raw materials in the production of wood products and are processed to produce biofuels.

Tree Trunks

The wood from the oil palm tree cannot be used directly as timber. The trunks of oil palm trees are also shredded to be added to natural fertiliser in plantations, or sold for RM8 to RM per strain. Additionally, the trunks can be processed to produce bioethanol, which in turn is used as biofuel.

Mesocarp Fibres (MF)

Mesocarp fibres are by-products from the extraction of crude palm oil (CPO) and palm kernel oil. The mesocarp fibres are also used as fuel in steam generation.

Palm Kernel Shells (PKS)

Like the mesocarp fibres, PKS are also the by-products of extracting crude palm oil (CPO) and palm kernel oil. It contains high calorific value which can be used as fuel to generate energy.

Palm Oil Mill Effluents (POME)

POME is the liquid waste or by-product generated from palm oil processing mills. Often regarded as pollutants, POME however can be processed and converted into biogas to generate renewable energy.

Oil Palm Biogas

Biogas from POME has the potential to be processed for other external applications such as the following:

- Use as biogas compressed natural gas (BioCNG), an alternative fuel for natural gas powered vehicles

- Feeding into natural gas pipeline or bottling and transporting for industrial use

- Supply energy to other operational units such as vapour absorption chiller, hot water and hot air production, biodiesel plant, EFB fibre plant as well as crushing plant

- Use as future second generation biofuel projects in the production of hydrogen, bio methane, and others.

Poplar

Hybrid poplars are commonly classified as short-rotation woody Biomass and can be grown on forest lands or on economically marginal crop lands.

Poplars (Populus spp.) are popular trees for landscape and agriculture use worldwide. They are known as "the trees of the people" and are considered one of the most important families of woody plants for human use. Poplars' incredibly fast growth has captured people's interest for many years. Around the world, people have used these trees for thousands of years to build homes, make tools and medicines, and protect river banks. They were also planted for windbreaks and shelterbelts. Poplars were first planted commercially in the Pacific Northwest in the late 1800s, and commercial tree farms expanded during the last 15 years for the pulp and paper industry.

Today, poplar uses are expanding to provide environmental benefits such as phytore-mediation, soil carbon sequestration, reduction in sediment run-off, improvement in soil quality, and habitat for wildlife. Poplars are also widely used for wood, veneer, and bioenergy. Researchers are working to improve poplars for bioenergy, carbon seques-tration, phytoremediation, and watershed protection, and some argue that poplars can be an important component of solving twenty-first century economic and environmen-tal problems as both human populations and greenhouse gases rise.

Current and Potential use as a Biofuel

In the late 1970s, hybrid poplar was part of the U.S. Department of Energy's (DOE) Bioenergy Feedstock Development Program. The primary target was fuel for cogen-eration of heat and electricity. The DOE and others are now interested in poplar for liquid fuels. Poplars can be farmed as short rotation woody crops (SRWC) and har-vested every two to five years. Other short rotation woody crops being considered for biofuel include willows and eucalyptus. Poplar trees, when intensively cultured, can produce substantial amounts of energy ranging from 7,735 to 8,634 Btu/lb (depend-ing on moisture content), which is equivalent to approximately 11 barrels of oil per acre per year.

Poplars are more desirable for biofuels than many other woody crops because of their fast growth, their ability to produce a significant amount of biomass in a short period of time, and their high cellulose and low lignin contents (Fig). For liquid fuels, the cellu-lose provides the carbohydrates to produce bioenergy and the low lignin content makes it easier to extract carbohydrates from the biomass. In addition, the development of poplar genotypes with improved yield, higher pest resistance, increased site adaptabil-ity, and easy vegetative propagation has made poplar a commercially valuable energy crop. The DOE also considers poplars to be one of the short rotation woody biomass crops that can be nationally developed. Poplars have some advantages over other bio-energy crops such as grasses because the wood does not need to be stored, which allows harvest to occur throughout the year.

General composition of poplar wood showing estimates of average cellulose and lignin content

Production and Agronomic Information

Poplars are genetically diverse and their ability to hybridize creates even more variation among the clones. To optimize biomass production levels, it is important to consider how different clones respond to different climatic factors and select the appropriate varieties for each region and sites within regions.

Poplar energy plantations will probably occupy idle, retired, or low-productivity cropland in order to avoid competition with food production. Careful attention to proper clonal selection, site preparation and weed management, fertility, and moisture will be critical to ensure success on these "marginal" sites. In drier regions, irrigation may be required.

Trees in poplar bioenergy farms can be planted at 1,500 trees per acre, which is much closer together than those intended for more traditional uses such as pulp and paper, which range from 34 to 360 trees per acre. This is done to speed the accumulation of biomass per acre rather than produce large individual trees. Studies are underway now to determine the economics of various planting systems. Depending on the management system being used, poplar energy farms may contain as few as 700 to as many as 5,700 trees per acre. Low-density tree farms are much less expensive to plant and produce fewer but larger stems, which may not be ready to harvest for eight to ten years. High-density plantings are expensive to plant but produce many small stems and more biomass that are ready to harvest within two or three years. Also of importance is the spacing between the stems. Optimal spacing studies are currently underway at poplar tree farms in the Pacific Northwest. Researchers are investigating optimal spacing for bioenergy, such as twin-row spacing with trees at three feet intervals within each row and ten feet between the twin rows.

Planting can be done with unrooted hardwood cuttings 0.5 to 3 feet long with viable buds or with rooted cuttings that are bare root or in containers Fig.). Many nurseries have hybrid poplar available to the public. The preferred cuttings or planting stock will be available from organizations researching the best clones for biofuels, including GreenWood Resources, Inc. in the Pacific Northwest and ArborGen in the southeastern U.S.

Planting of hybrid poplar with cuttings occurs in the early spring with buds placed upright

In early spring after the ground has thawed, unrooted cuttings are generally planted manually. Efficient mechanical planting methods are also being developed. In general, three quarters of the cutting should be placed in the soil with at least two inches above ground. It is important to place cuttings in the soil with the buds pointing upright. For longer cuttings, more of the cutting may be above the surface to promote the growth of multiple shoots. The unrooted cuttings develop adventitious roots and the first leaves appear within a few weeks of planting.

Since poplar grows well in poor soils, fertilizer inputs are low compared to other crops. Nutrient inputs need to be based on local soil analysis. Nitrogen fertilizer application has not produced significant yield increases in preliminary Lake States' trials. Weed control should only be needed for one or two years until the poplars form a closed canopy. Pesticides may be needed to inhibit cottonwood leaf beetle and other pests. There are several common damaging or fatal diseases, including leaf rusts and stem cankers such as Septoria musiva. Planting pest-resistant clones is the most effective way to avoid losses from pests. Fortunately, pest resistance can be bred into poplar clones, but there are strong regional differences in response. In the Mississippi River Valley, for example, native eastern cottonwoods are less susceptible to disease than the hybrid poplars. In the Pacific Northwest, poplar hybrids with a P. trichocarpa parent do well but they tend to be highly sensitive to the diseases that are present in the Lake States region.

Poplar utilized for biofuels will be grown as SRWC. The rotation cycles will be much shorter than traditional poplar plantations. The length of the cycle will vary in different regions, with longer ones in more northern latitudes and shorter ones in more southern latitudes. The poplar biomass can be harvested again in another two to three years as sprouts emerge from cut root stalks (coppice), forming new stands of poplar. This pattern can be repeated several times before replanting is required.

Harvest

Harvest of hybrid poplars with a modified forage harvester.
With the single pass system, the chips can be directly loaded into a trailer or truck

Harvesting of poplars grown on six- to ten-year rotations can be done with tradition-al timber harvesting methods if individual trees are large enough. The trees are then bunched and the wood is chipped for biofuels.

For small trees grown on two- to three-year rotations, a fully mechanized New Holland harvester may be a more economically attractive option. A modified forage harvester cuts and chips the trees as it moves along the row.

After harvest, the chips are ideally shipped directly to the biorefinery. The moisture content at harvest ranges from 40-58%, and the biorefinery is not expected to need dry chips. If the biorefinery is not able to take the chips, they may be stored for the short term where they were harvested.

Potential Yields

Poplar yield depends on climate, site quality, clone, age, spacing, and silvicultural con-ditions and treatments. In general, yield is lower in unirrigated and unfertilized tree farms in the upper Midwest and higher in the Southeast where growing seasons are long and the Pacific Northwest where irrigation and soil moisture may be more abun-dant. Yield estimates range from 1.25 to 8.61 dry tons per acre per year (Table). Ge-neticists are working on increasing the biomass yield, adaptability, and pest resistance of poplar. To increase yield over time, it will be important to select the best clones for particular sites within each region.

Region	Production Estimates (Dry ton acre^{-1}yr^{-1})	Growth Cycle (yrs)
Lake states	1.25-3.35	5 to 8
Upper Midwest	2.45-5.13	13
Mississippi River Valley	2.01-2.99	8 to 10
Pacific Northwest	3.08-8.61	6 to 11

Production Challenges

For poplars to become a successful biofuel crop, there must be a suitable and available land base in close proximity to potential biorefineries. Because poplar may be targeted for marginal and idle lands, there may be difficulty in obtaining water for irrigation. For small landowners to be successful at growing poplar, they may need to form co-ops and obtain specialized technical support. Local communities will also need to accept poplar plantations and biorefineries. The larger political, social, and economic forces, espe-cially the price of a barrel of oil compared to liquid biofuels, will be among the greatest challenges to overcome.

If the biorefinery can utilize poplar chips that also contain leaves and bark, harvest can occur during the growing season as well as the dormant season. Research is underway to investigate the conversion technology of producing biofuels from clean poplar chips compared to chips containing leaves and bark. Some conversion processes may only be able to utilize clean poplar chips, which will require that the trees be grown on longer rotations and harvested with more traditional timber methods to remove the bark before being sent to the biorefinery.

Estimated Production Cost

Landowners will naturally decide which crops they will grow on their land. Understanding the costs and returns from poplar energy crops will allow growers to consider this alternative crop and decide where it would best be grown on their fields.

Similar to other agricultural row crops, poplar production costs may include land rental, site preparation, planting material (cuttings), planting, weed and pest control, fertilizer, irrigation, root stock removal, labor and management, crop insurance, harvesting, and transportation. However, many of these cost factors will differ widely across regions and scenarios regarding the ease of establishing and harvesting poplars. For the best estimate, potential growers will need to gather key information on current farming expenditures of other traditional crops. Land cost is a significant part of the cost of producing poplar as an energy crop, and utilizing marginal or less productive cropland may significantly reduce the cost compared to other productive lands. When planted on marginal land, the break-even prices of biomass feedstock are most sensitive to changes in biomass yield and harvest costs.

Current research on growing hybrid poplar as an SRWC for biofuel production shows great promise to increase profitability. Cash flow models of production costs and expected yields from the billion ton update show that poplar biomass ranges from $25 to $60 per dry ton. This estimate is comparable to other dedicated biomass production systems and does not include the cost of transportation from the field to the biorefinery. A preliminary study of short-rotation poplars in Michigan showed 4 to 5 dry tons/acre yield where the break-even price of poplar biomass was estimated to be $108 (delivered to the mill) based on $60/ton. The preliminary study suggests 60% yield gains in poplar biomass per acre to compete with production cost of corn-based biofuels.

Although the SRWC approach is completely different than long-rotation (10 to 20 years) poplars from the pulpwood industry, there are lessons that could be learned from the previous studies on long-rotation poplar production costs and returns. For example, in the Southeast, Gallagher et al. estimated that delivered break-even costs of poplar biomass to a pulp mill (assumed transport distance of 31 miles) are between $75.58 to $89.28/dry ton. More research and improvements in poplar yield and production costs may make poplar biofuels more feasible.

Environmental/Sustainability Issues

Poplar trees provide numerous environmental benefits including conservation of soil, water, and biodiversity. Studies show that short rotation poplars reduce soil erosion and surface runoff. Poplar trees can also be used to remove contaminants from municipal biosolids and for phytoremediation. In agriculture areas, short rotation poplars can reduce stream flow rates and the amount of nitrogen and sediments entering streams. In terms of biodiversity, research in the northern plains and Midwest found an increase in bird diversity in poplar plantings compared with traditional row crops.

Theoretical life cycle assessments (LCAs) of biofuels made from hybrid poplars show their potential to substantially reduce greenhouse gas (GHG) emission compared with traditional fossil fuels. In fact, short rotation poplar may actually be carbon negative, sequestering more carbon than is released. Poplar may be better in this respect than alternative crops like switchgrass and corn-soybean rotations. It is important to note that these LCAs depend largely on assumptions about many things, including the cropping system used, the previous land use, the location of the crop, the biofuel conversion system employed, and the fuel end use. Because commercial production of both hybrid poplar energy crops and bio-based fuels is in its infancy, real-world examples on which LCAs can be conducted are generally lacking. Even so, the preliminary LCAs that have been done suggest key areas where cropping systems and fuel production systems might be improved to further improve GHG balances. Key among these are reduction of fertilizer, water, and fossil fuel inputs per unit of biomass grown and transported.

There are environmental concerns with producing poplar-based biofuels, including impacts on water and air quality such as isoprene release during crop growth. Other concerns being examined include GHG debts incurred and nitrogen mobilized during land use change. Improvements in production systems should eventually be able to ameliorate these environmental concerns.

Feasibility

Fast growth and wide site adaptability have made poplar a suitable tree species to grow for multiple purposes including biofuels production. In some locations, they also are known for being useless when markets disappear. For poplars to be a viable commercial scale feedstock for biofuels, stable markets must be established and contracts must be made so the burden is not solely on the grower. Co-ops for growers may make this more feasible.

As a biofuel crop/feedstock, poplars may have an advantage over other feedstocks because poplars can potentially be harvested in different seasons and thus provide a continuous supply of feedstock to the biorefinery. Poplars also produce higher amounts of energy than other feedstocks and are predicted to displace more gasoline and diesel than corn, soybeans, reed canary grass, and switchgrass. The amount of energy produced does depend on poplar management and production input. Moreover, compared to other wood sources, it is much easier to convert hybrid poplar into liquid biofuels. The high genetic variation among poplars and many desirable traits provide an opportunity to develop them as ideal feedstock for biofuels.

Pine

Biomass, such as pine wood, is a very common and generally accessible source of renewable energy. Traditionally, biomass has been utilized by direct combustion; however, the development of a biomass thermal conversion technology represents a promising alternative for energy production.

Non-woody Biomass

The various agricultural crop residues resulting after harvest, organic fraction of municipal solid wastes, manure from confined livestock and poultry operations constitute non-woody biomass. Non-woody biomass is characterized by lower bulk density, higher void age, higher ash content, higher moisture content and lower calorific value. Because of the various associated drawbacks, their costs are lesser and sometimes even negative.

Proximate Analysis of Presently Selected Non-Woody Biomass Plant Components And Coal Biomass Mixed Briquettes

Freshly chopped non-woody biomass components have a large amount of free moisture, which must be removed to decrease the transportation cost and increase the calorific value. In the plant species selected for the present study, the time required to bring their moisture contents into equilibrium with that of atmosphere was found to be in the range of 15 to 20 days during the summer season (temperature :35-45°C and moisture: 6-14%).

The studies of the proximate analysis of fuels /energy sources are important because they give an approximate idea about the energy values and extent of pollutants emissions during combustion. The proximate analysis of different components of Gulmohar and Cassia Tora plant and these biomass species component briquettes with coal are presented in below tables. The data for proximate analysis of the components of these species are very close to each other and hence it is very difficult to draw a concrete conclusion. However, it appears from these tables that Cassia Tora biomass species has somewhat higher ash and lower fixed carbon contents than these of Gulmohar biomass species and the ash contents being more and volatile matter is less when 95% coal mixing with 5% biomass and 90% coal mixing with 10% biomass but when 85% coal mixing with 15% biomass and 80% coal mixing with 20% biomass then ash content is being less and volatile matter is more.

Calorific Values of Presently Selected Non-Woody Biomass Plant Components

The calorific values of the fuels/energy source are important norms for judging its quality to be used in electricity generation in power plants. It provides an idea about the energy value of the fuel and the amount of electricity generation.

Calorific values data listed in the first two tables below, indicate that among all the studied biomass species, calorific values of wood component of both biomasses have higher in comparison to leaf and nascent branch. Gulmohar biomass species were found to be little bit higher than that of Cassia Tora biomass. Two tables below shows that calorific value of coal mixed Gulmohar biomass (different component in different ratio) were found to be higher than that of coal mixed cassia tora biomass (different component in different ratio).

Amongst the four different ratios, ratio 80:20 gives the highest energy value in all mixed component and 85:15 also gives higher energy value except leaf component of both biomass in respect to other two ratios (95:05 and 90:10).

Comparison of data listed in first three tables shows that in difference to coals included in the present study, both non-woody biomass materials have considerably higher calorific values and very lower ash contents. Fourth and fifth table indicates that calorific values of biomass species are something lower but ash content are also lower in compare to coal. This is definitely an benefit over fossil fuels. It is thus clear that these non-woody biomass resources will result in higher power production in the plant with slight emission of suspended particulate matters (SPM).

Table: Proximate Analysis of Gulmohar (Local name: Krishnachura).

Component	Proximate Analysis (Wt. %, air-dried basis)				Gross Calorific Value (Kcal/ kg, Dried Basis)
	Moisture	Ash	Volatile Matter	Fixed Carbon	
Wood	9.00	3.00	72.68	15.00	4549
Leaf	8.90	7.20	70.11	15.00	3947
Nascent branch	9.80	4.20	70.05	14.22	4061

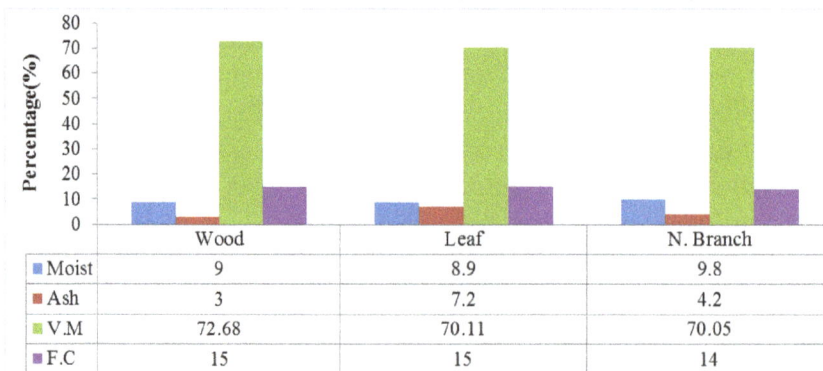

	Wood	Leaf	N. Branch
Moist	9	8.9	9.8
Ash	3	7.2	4.2
V.M	72.68	70.11	70.05
F.C	15	15	14

Figure: Variation of Proximate Analysis of Gulmohar Biomass

Table: Proximate Analysis of Cassia Tora (local name: Chakunda).

Component	Proximate Analysis (Wt. %, air-dried basis)				Calorific Value (Kcal/ kg, Dried Basis)
	Moisture	Ash	Volatile Matter	Fixed Carbon	
Wood	11.00	7.80	68.50	12.00	4344
Leaf	11.50	7.40	69.00	14.00	4113
Nascent branch	10.00	5.20	70.00	14.00	3697

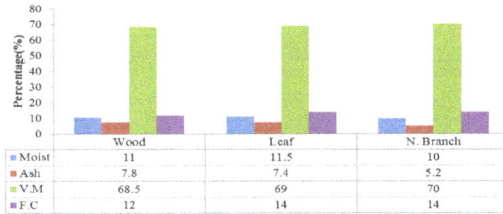

	Wood	Leaf	N. Branch
Moist	11	11.5	10
Ash	7.8	7.4	5.2
V.M	68.5	69	70
F.C	12	14	14

Figure: Variation of Proximate Analysis of Cassia Tora Biomass

Table: Proximate Analysis of Non-coking coal.

Component	Proximate Analysis (Wt. %, air-dried basis)				Calorific Value (Kcal/ kg, Dried Basis)
	Moisture	Ash	Volatile Matter	Fixed Carbon	
Lingaraj Mines	8.90	41.20	21.70	29	4237

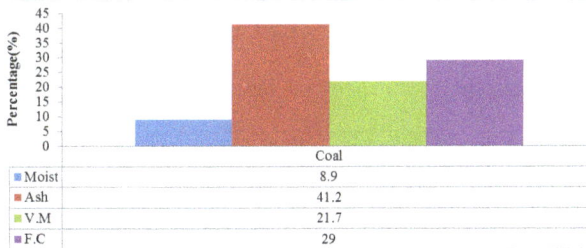

	Coal
Moist	8.9
Ash	41.2
V.M	21.7
F.C	29

Figure: Variation of Proximate Analysis of Non-coking Coal

Table: Coal- Gulmohar Biomass Different Component.

Ratio (Coal: Biomass)	Proximate Analysis (Wt. %, Air Dried Basis)				Calorific value (Kcal/ kg, Dried Basis)
	Moisture	Ash	Volatile Matter	Fixed Carbon	

Main Wood

95:05	7	36	25	32	3214
90:10	5	34	31	30	3497
85:15	4	36	33	27	3748
80:20	4	34	33	29	4087

Leaf

95:05	4	35	29	32	3422
90:10	4	36	31	29	3483
85:15	5	29	35	31	3077
80:20	6	31	33	30	3830

Nascent Branch

95:05	4	37	32	27	3584
90:10	3	33	35	29	3551
85:15	6	29	39	26	3557
80:20	7	30	42	21	3801

Table: Coal- Gulmohar Biomass Different Component

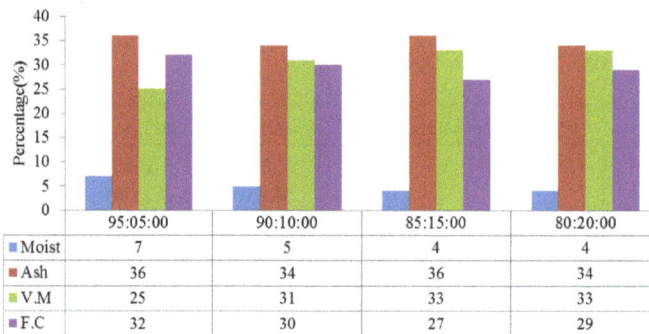

	95:05:00	90:10:00	85:15:00	80:20:00
Moist	7	5	4	4
Ash	36	34	36	34
V.M	25	31	33	33
F.C	32	30	27	29

Figure: Variation of Mixture of Coal and Gulmohar Biomass (Wood)

Ratio (Coal: Biomass)	Proximate Analysis (Wt. %, Air Dried Basis)				Calorific value (Kcal/ kg, Dried Basis)
	Moisture	Ash	Volatile Matter	Fixed Carbon	

Main wood

95:05	3	36	36	25	3146
90:10	4	36	33	27	2980
85:15	4	37	39	20	3482
80:20	6	35	41	18	3454

Leaf

95:05	3	39	29	29	3275
90:10	4	39	29	28	3668
85:15	4	31	39	26	3051
80:20	4	33	34	29	4143

Nascent Branch

95:05	4	39	32	25	3471
90:10	7	37	29	27	3211
85:15	3	31	39	27	3675
80:20	3	36	39	22	3672

Table: Coal: Cassia Tora Biomass Different Component

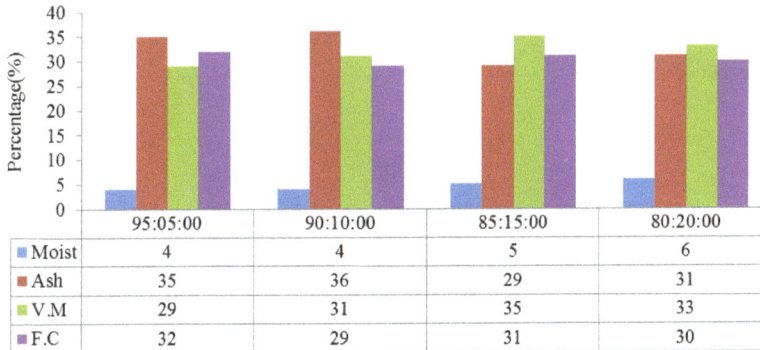

	95:05:00	90:10:00	85:15:00	80:20:00
Moist	4	4	5	6
Ash	35	36	29	31
V.M	29	31	35	33
F.C	32	29	31	30

Figure: Variation of Mixture of Coal and Gulmohar Biomass (Leaf)

Ash Fusion Temperature of Presently Studied Non-Woody Biomass Species

It also experimentally finds out the ash fusion temperatures to confirm its safe operation in the boiler. Ash fusion temperature of solid fuel is an important parameter affecting the operating temperature of boilers. Clinker creation in the boiler usually occurs due to low ash fusion temperature and this hampers the operation of the boiler. Hence the study of the ash fusion temperature of solid fuel is essential before its operation in the boiler. The four characteristic ash fusion temperatures were identified as: (i) initial deformation temperature (IDT) – first sign of change in shape; (ii) softening temperature (ST) – rounding of the corners of the cube and shrinkage; (iii) hemispherical temperature (HT) – deformation of cube to a hemispherical shape; and (iv) fluid temperature (FT) – flow of the fused mass in a nearly flat layer. Identical shapes at these temperatures were obtained for all the studied non-woody biomass species like Gulmohar, cassia tora and coal mixed these biomass. Data for the ash fusion temperatures (IDT, ST, HT and FT) for have been listed in table below.

Biomass Species / Coal-Biomass Mixed Ratio	Ash Fusion Temperatures(°C)			
	IDT	ST	HT	FT
Cassia Tora	893	1045	>1400	>1400
Gulmohar	1058	1249	>1400	>1400

Coal : Biomass (90:10)	1160	1297	>1400	>1400
Coal : Biomass (80:20)	1188	1298	>1400	>1400

Table: Ash Fusion Temperatures of Selected Biomass Species and Coal- Biomass Mixed Sample

IDT: Initial Deformation Temperature
ST: Softening Temperature
HT: Hemispherical Temperature
FT: Flow Temperature

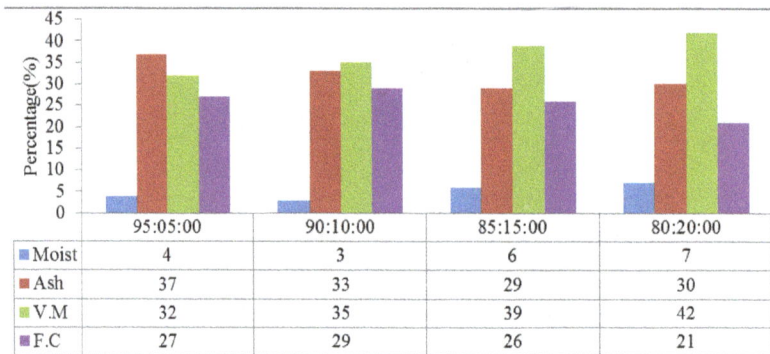

	95:05:00	90:10:00	85:15:00	80:20:00
Moist	4	3	6	7
Ash	37	33	29	30
V.M	32	35	39	42
F.C	27	29	26	21

Figure: Variation of Mixture of Coal and Gulmohar Biomass (Nacent Branch)

Algae

Algae have certain qualities that make the organism an attractive option for biodiesel production. Unlike corn-based biodiesel which competes with food crops for land resources, algae-based production methods, such as algae ponds or photobioreactors, would "complement, rather than compete" with other biomass-based fuels. Unlike corn or other biodiesel crops, algae do not require significant inputs of carbon intensive fertilizers. Some algae species can even grow in waters that contain a large amount of salt, which means that algae-based fuel production need not place a large burden on freshwater supplies.

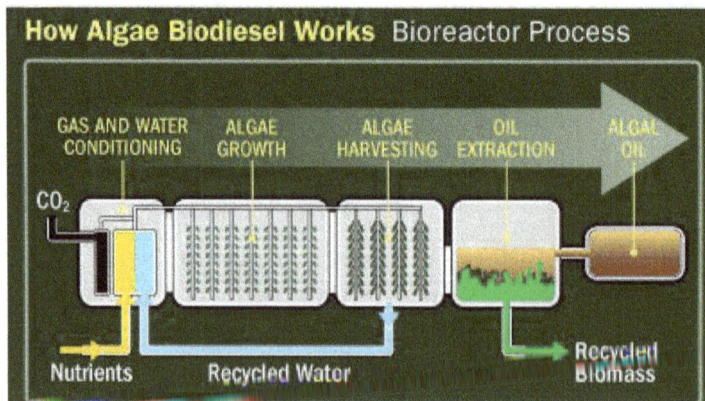

Several companies and government agencies are funding efforts to reduce capital and operating costs and make algae fuel production commercially viable. Companies such as Sapphire Energy and Bio Solar Cellsare using genetic engineering to make algae fuel production more efficient. According to Klein Lankhorst of Bio Solar Cells, genetic engineering could vastly improve algae fuel efficiency as algae can be modified to only build short carbon chains instead of long chains of carbohydrates.

Algae Farming for Biofuel Production

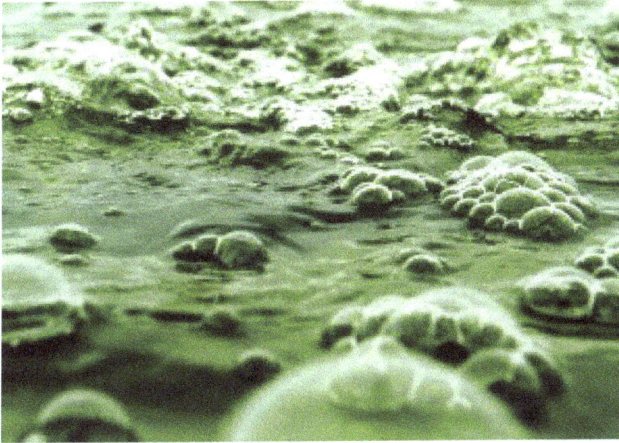

High oil prices, competing demands between foods and other biofuel sources, and the world food crisis, have ignited interest in algaculture (farming algae) for making vegetable oil, biodiesel, bioethanol, biogasoline, biomethanol, biobutanol and other biofuels, using land that is not suitable for agriculture. Algae holds enormous potential to provide a non-food, high-yield, non-arable land use source of biodiesel, ethanol and hydrogen fuels. Microalgae are the fastest growing photosynthesizing organism capable of completing an entire growing cycle every few days. Up to 50% of algae's weight is comprised of oil, compared with, for example, oil palm which yields just about 20% of its weight in oil.

Algaculture (farming of algae) can be a route to making vegetable oils, biodiesel, bioethanol and other biofuels. Microalgae are one-celled, photosynthetic microorganisms that are abundant in fresh water, brackish water, and marine environments everywhere on earth. The potential for commercial algae production is expected to come from growth in translucent tubes or containers called photo bioreactors or open ocean algae bloom harvesting. The other advantages of algal systems include:

- Carbon capture from smokestacks to increase algae growth rates
- Processing of algae biomass through gasification to create syngas
- Growing carbohydrate rich algae strains for cellulosic ethanol
- Using waste streams from municipalities as water sources.

Corn

The biggest biofuel in the United States right now, corn sometimes gets a bad wrap. Corn ethanol is more sustainable than petroleum, but it has been a centerpiece for debates on using agricultural crops for fuel. It's true that corn used for fuel is corn that could have been someone's dinner, but Runge explained that even once the corn oil has been extracted for ethanol, there is still a byproduct of distiller corn that can be fed to animals. "It is taking food and putting it into fuel, but there is a byproduct of doing that that still can go to animals. It's not one to one, but it's not all bad," Runge said. Even given the possible by-products it seems that corn is, at best, a short-term solution. Much like sugar cane, corn is one of the best options we have available now, but because the process is expensive and has high energy consumption rates, Runge felt it should be high on the list, but may fade from use over time.

Corn Stover

Corn stover refers to stalks, leaves and cobs that remain in fields after the corn harvest. This biomass can be used in producing ethanol. Corn stover is the primary biomass source being used for producing cellulosic ethanol.

The corn stover ethanol byproduct has three times the concentration of nitrogen as the original cornstalks.

Corn stover is the largest quantity of biomass residue in the United States. Around 120 million tons of biomass residue is available annually. This stover has the potential of supplying 23 to 53 billion liters of fuel ethanol.

The two important routes to converting corn stover into biofuels are: biological conversion and thermochemical conversion. Biological conversion involves four steps: pretreatment, enzymatic saccharification, fermentation and recovery. The thermochemical conversion involves gasification and pyrolysis. Pyrolysis oil (bio-oil) is produced when corn stover is rapidly heated in the absence of air to temperatures ranging from 400 to 600 degrees C. Bio-oil is composed of hydrocarbons, gases and charcoal. Bio-oil can be refined into biofuel. In gasification, the stover is gasified and carbon monoxide, hydrogen and carbon dioxide in the synthesis gas are fermented into ethanol.

The height at which corn has to be harvested to ensure the most economical and efficient stover harvest is an important factor in ethanol production. Research indicates that a normal-cut harvest results in the most economical and efficient stover harvest for biofuel production. At least 16 inches of stubble should remain on the field for it to be considered normal-cut stover harvest.

Grasses

Grass pellets have great potential as a low-tech, small-scale, environmentally-friendly, renewable energy system that can be locally produced, locally processed and locally consumed. As the US focuses on energy security, grass bioenergy is one of the ways that rural communities can move towards energy security. New York State has about 1.5 million acres of unused or underutilized agricultural land, most of which is already growing grass.

Grass biofuel production does not need to divert any of the current agricultural productivity into the energy market; this biomass industry can be completely independent from, but complimentary to, the production of food or animal feed. It is also a very "farmer-friendly" way to get producers exposed to biofuel production. Some research and development is needed to optimize stoves and boilers for grass combustion, and to minimize emissions.

Benefits of using Grass for Energy

Perennial grasses have many benefits as a bioenergy crop. The simplest way to think of grass is as an efficient and fast growing solar energy collector that is relatively easy to grow, harvest, and process. Grasses not only sequester and store vast amounts of carbon in the root systems and soil, but conveniently occur globally in a wide range of geographies, climates, and soil types.

Grasses can be grown on marginal lands ill suited for continuous row crop production and/or in open rural land currently not in agricultural production. They yield more biomass per acre, and, once established, require far fewer inputs in comparison to annual crops that require more diesel, fertilizer, and pesticides.

Additionally, perennial grasses grown for energy can provide a new revenue stream and profit center to farmers and other landowners, and afford important water quality and wildlife benefits. Grasses and other agriculturally produced crops can be grown easily (with conventional equipment), quickly, and in large acreages and volumes. This can help increase the production of biomass fuels by utilizing local resources. Soil erosion, water quality, and wildlife benefits can also be enhanced depending on what type of land and current crop cover is converted to energy crops.

Energy studies indicate that significant gains in energy return and reducing carbon emissions can be achieved with using Switchgrass as a biomass fuel. Switchgrass used for heating has an energy output to input ratio of at least 10 to 1, compared to other bio-energy sources with output to input ratios around 1 to 1. A recent study determined that one acre of farmland is capable of producing an average annual yield of herbaceous biomass sufficient to meet the annual space- and water-heating needs of an average home. An existing energy prospect, with planning and conservation, is the ability for communities to produce their own heating fuel through local farmers growing grasses and farm supply cooperatives (or other aggregating businesses) densifying and delivering fuel.

The most promising areas for development of a grass-based energy industry are the north central and northeastern regions of the United States, where there is sufficient agricultural land base and heating costs are high due to long winters. As an alternative fuel for heating, grasses have Btu levels approximately 95 percent of wood. Densified grass fuel is competitive in price with fuel oil, natural gas, propane, and electricity.

Choice Grasses for Fuel

No one grass species can be grown effectively in all regions and climates, however, the most broadly considered grasses for energy production are Switchgrass (and other native prairie grasses such as Big Bluestem and Prairie Cordgrass). Miscanthus, a super high-yielding crop, has garnered much interest and is now being studied. Reed Canarygrass is often naturally present and high yielding in wet, marginal areas, however, it is also recognized as invasive—choking out other native wetland species—so its use as an energy crop is more contentious. Each has its own benefits and disadvantages as a biomass fuel source.

When considering which is the best choice, the first consideration is generally the yield per acre in any given microclimate/soil type, as this greatly influences the economics of conversion of the crop to a useful form for energy extraction. Another consideration is the mineral/ash content of a given grass on a given plot, which may affect the value of the crop as a densified fuel for thermal applications. Another consideration may be harvest windows as influenced by local climates. What will the moisture content of the harvested grass be? Will this limit uses? Are there other users, birds, for example, of the grass as it is growing and how do they shape options for harvesting?

Below are three major fuel grasses and a brief description of each.

- Switchgrass (Panicum virgatum) is native to the United States and one of the best herbaceous (not woody) energy crops because of its perennial growth habit, high yield potential on a wide variety of soil conditions and types, compatibility with conventional farming practices, and value in improving soil and water conservation and quality.

- Miscanthus (Miscanthus x giganteus, a natural hybrid of Miscanthus sinensis and M. sacchariflorus) is a giant, perennial warm-season grass native to Asia that is generating much enthusiasm for extremely high yields and very high cold tolerance. Miscanthus, however, does not produce viable seed (a sterile hybrid) so must be propagated by planting underground stems, called rhizomes.

- Reed Canarygrass (Phylaris arundinacea) is a perennial wetland grass, native to parts of the US, Europe, and Asia. It is a cool-season grass that is less productive than warm-season grasses. It is winter hardy so can be grown in colder climates and under shorter growing seasons; however, many ecologists and conservation departments consider it an invasive species because it frequently out competes and threatens natural wetland species.

Managing Grass for Energy

In general, grasses grown for energy are managed for biomass yield rather than forage or nutritive quality. In fact, lower nutrient levels (nitrogen, sulfur, chlorine, etc.) may improve fuel quality and reduce emissions. As expected, the growth and yield of the grass crop is highly dependent on soil conditions, moisture, fertility, weed control, and timing of harvest. During the growing season, modest use of fertilizers may be needed to maintain soil fertility and improve crop yields. Careful attention must be paid to ensure that crops are not overfertilized for risk of leaching surplus nutrients into ground and surface waterways.

Native, warm-season grasses like Switchgrass are widely adaptive once established, however, require attentive weed control in the first year of establishment so cool season grasses do not overwhelm it. Nitrogen fertilizer is not recommended in the first year to reduce competition from grassy weeds. Switchgrass should be harvested once per year, generally after frost, using standard haying equipment. Grasses cut in the fall and left to over-winter are far lower in yield but have been shown to leach out potassium and chlorine, two minerals that may create issues during combustion.

Miscanthus can be tricky to establish and should be harvested late in the fall, after senescence, using standard farm equipment (i.e,. corn silage choppers, balers).

Each species has its own management protocol, and state university extension offices can provide information on specific seeding, harvesting, fertilizing, and weed-management practices.

Densifying Grasses

To utilize the energy stored in grass, the crop must be harvested and processed into a 'user-friendly' format, either on farm or transported to an off-farm facility.

Like wood, grasses can be densified into a durable, high-quality pellet fuel, sized to a standard ¼ -inch diameter. There are also prototypes of 'portable' pelletizers that could support a distributed production model, although, to date, these units are not commercially available and the fuel does not match the quality of those from centralized pellet plants.

A less-expensive densification method (higher throughput per hour) is by forming the grass into larger briquettes, also called tablets or pucks or cubes, allowing the material to be handled and stored easily, transported economically, and burned efficiently. One advantage to this method is that grasses dry in the field, reducing drying costs at a pellet mill. While this process is easier than making ¼-inch pellets, so far there are very few systems that can successfully burn these larger forms of grass-based fuels, and they may not be as durable or flowable through augers and upon conveyers.

Grass Combustion

Grasses have 95 percent of the Btu value of wood and several pioneering companies are beginning to produce high-quality grass pellets for heating. Historically, since biomass combustion systems were designed around wood, simply substituting grass for wood in the same combustion system will generally not produce satisfactory results.

Grasses have a higher ash content and a different chemical composition, therefore distinct combustion systems are needed to handle these differences. During combustion, higher chlorine and potassium levels in grasses vaporize and form salts on boiler walls. These salts can cause 'clinkers' (incombustible residues) in systems not specifically designed to handle grasses, reducing performance markedly.

At both the commercial and residential scale, there is a growing number of equipment manufacturers producing multi-fuel combustion systems that show promise, e.g., 80-90 percent efficient "close-coupled" gasifier pellet stoves and multi-fuel stoves and furnaces capable of burning moderately high ash pelleted fuels.

Soybean

Soy has been a popular biofuel for several years now. In a process called trans-sterification, producers squeeze the oil from seeds and use it in products such as biodiesel and jet fuel. It is a relatively easy and inexpensive rendering process As is the case with many agricultural crops, there is debate over the extent to which soy could be utilized. Crops like soybeans are dietary staples to many people, and researchers are reluctant to rely too heavily on traditional food crops as fuel sources. The seed oil that goes in a gas tank could have gone to someone's stomach, and it may prove difficult to say one destination is more valuable than the other. While soy is used widely, it is not as popular as corn or sugar cane, nor does it have the amount of resources that cellulose or algae provide. This makes it a short-term solution that deserves attention, but not a higher spot on the list.

Sugar Cane

In the world of biofuel production today, sugar cane is second to corn as the most widely used. Sugar cane grows in warm parts of the world in abundance, and has helped countries like Brazil to become energy independent. Sugar cane ethanol isn't the easiest on a car's internal workings as it can gum up the engines of older cars, but with flex-fuel cars and gasoline blends varying from 20 percent ethanol all the way to 100 percent ethanol, Brazil shows it can work. Sugar cane uses more of the plant than seed-based fuels like corn and soy, but it still does not utilize all of the plant. Also, because it can only be grown in the tropics, there are limits to how much sugar cane can be grown. Still, because sugar cane is a developed technology that is already in wide spread use, it earns a higher place on the list.

References

- What-is-biomass, renewable-energy: reenergyholdings.com, Retrieved 15 July 2018

- Woody-biomass-resources: alternative-energy-news.info, Retrieved 22 May 2018

- 10-biomass-advantages-disadvantages: visionlaunch.com, Retrieved 12 March 2018

- Coconut-biomass: bioenergyconsult.com, Retrieved 29 April 2018

- Poplar-populus-spp-trees-for-biofuel-production-70456: extension.org, Retrieved 30 April 2018

- What-is-algal-biomass: bioenergyconsult.com, Retrieved 22 May 2018

- Corn-stover, renewable-energy: agmrc.org, Retrieved 31 March 2018

- Top-10-sources-for-biofuel-1769457447: seeker.com, Retrieved 28 April 2018

Chapter 4

Sources of Biofuels

Science and technology has undergone tremendous development in the past decade, which has resulted in the production of various biofuels. Some of the common sources of biofuels are corn, sugarcane, cellulose, Camelina, Jatropha, rapeseed, algal oil, etc. which have been extensively examined in this chapter.

Corn

Corn (Zea mays) is a popular feedstock for ethanol production in the United States due to its abundance and relative ease of conversion to ethyl alcohol (ethanol). Corn and other high-starch grains have been converted into ethanol for thousands of years, yet only in the past century has its use as fuel greatly expanded. Conversion includes grinding, cooking with enzymes, fermentation with yeast, and distillation to remove water. For fuel ethanol, two more steps are included: using a molecular sieve to remove the last of the water and denaturing to make the ethanol undrinkable.

Current Potential for use as a Biofuel

Corn grain makes a good biofuel feedstock due to its starch content and its comparatively easy conversion to ethanol. Infrastructure to plant, harvest, and store corn in mass quantities benefits the corn ethanol industry. Unlike sugarcane, in which squeezed sugar water can be directly fermented, corn starch must be cooked with alpha and gluco-amylase enzymes to convert the starch to simple sugars. Cellulosic feedstocks are even more recalcitrant and require time and energy to convert to simple sugars. Under the renewable fuel standard set by Congress in 2007 (RFS-2), grain-based ethanol can make up 15 billion gallons of the 36 billion gallon-per-year requirement. Corn-based ethanol production capacity in 2009 was 10.6 billion gallons. The addition of idled capacity would increase potential production to 12.5 billion gallons per year (Renewable Fuels Association).

Corn production in the United States reached record highs in 2009 with 13.2 billion bushels from 86.5 million acres (National Ag Statistics Service). Using the current corn-to-ethanol conversion of 2.8 gallons of ethanol from a bushel of corn, total U.S. corn production could result in approximately 37 billion gallons of ethanol, which would provide approximately 26% of our 137 billion gallon-per-year gasoline consumption (Energy Information Administration). However, using all of our corn for ethanol is neither realistic nor necessary and has not been proposed. Creating the 15 billion gallons required under the RFS-2 would call for 5.4 billion bushels or about 41% of our 2009 corn crop. Although this percentage seems rather high, one-third of the weight and 100% of the nutritional content of corn entering an ethanol dry mill biorefinery is returned to the feed market as distillers grains. These distillers grains can be used to replace corn in the diets of cattle, swine, and poultry. When this replacement is calculated into the overall consumption figures, it lowers the number to 27% of our 2009 corn crop, or only 3.6 billion bushels of corn to produce the 15 billion required gallons. Throughout history, the United States has seen a steady increase in the yields of both corn and ethanol production. It is very likely that the United States will be able to increase corn ethanol production without expanding to new acres and still have plenty of corn remaining to meet other domestic use and export demands.

One bushel of corn grain will yield one-third ethanol, one-third distillers grains, and one-third carbon dioxide, or 17 pounds of distillers grains and 2.8 gallons of ethanol.

Biology and Adaptation

Corn (Zea mays) originated in Central America with the first domestication, purported to be in the Tehuacan Valley of Mexico. Spreading throughout the North American continent, corn became an important crop for early Americans. At its peak in 1917, 111 million acres of corn were planted in the United States. Today corn is planted on every continent in the world except Antarctica and is grown throughout many states in the United States, ranging from southern North Dakota to Texas and eastward to New York. Corn is well adapted to growing in temperatures between 50° and 86°Fahrenheit. To produce grain, corn will use approximately 22 to 28 inches of water, which requires 12 to 20 inches of rainfall or irrigation during the growing season. Despite popular misconceptions, nearly 90% of U.S.-grown corn is fed by natural rainfall only, with no irrigation necessary. Many parts of the upper Midwest are well suited to grow corn, and this area is sometimes referred to as the Corn Belt.

Production and Agronomic Information

Corn in the upper Midwest is seeded between March and May and harvested between

September and November in most years. A majority of corn planted today has genetic resistance to some weeds, insects, and plant pathogens. Corn hybrid resistance to various pests and pathogens is a result of biotechnology and plant breeding. Biotechnology traits aid producers in the control of weeds and insects, greatly reducing the amount of pesticides entering the environment. Much of the Corn Belt rotates with other crops, such as soybeans or wheat, to break weed, insect, and disease cycles as well as to reduce the cost of production. Corn responds best to highly fertile soils with supplemental fertilizer applied in most years. Fertilizer may be inorganic chemical fertilizer or manure.

Major nutrients required by corn are nitrogen, phosphorus, and potassium. Inorganic nitrogen fertilizer production is very energy-intensive and as a result, nitrogen fertilizer represents nearly 30% of the energy inputs in corn production. Other major inputs include diesel fuel for tractors, transportation, irrigation, and electricity for irrigation and grain storage.

Potential Yields

The average national corn yield was 165 bushels per acre in 2009. Corn yields have increased by approximately 2 bushels per acre each year since 1940. This increase will likely continue into the future, with some predicting the yield trend will amplify at a greater rate due to biotechnology and advancements in breeding. Ethanol yield per acre would be 462 gallons per acre from corn yields of 165 bushels per acre. An acre of sugarcane can produce an approximate 35 ton yield, resulting in about 560 gallons of sugarcane ethanol.

Production Challenges

Corn production has been blessed with nearly 100 years of infrastructure build-up and research. Farmers have great knowledge and experience in growing corn. This infrastructure and grower intelligence make corn a natural crop for expanded uses such as ethanol. Yet high production costs and high inputs make corn a very intensive crop. Other bioenergy crops may be less intensive and require fewer inputs. The cost versus profit per acre needs to be compared, as economics is a major driver in deciding which crop is best. Growing another crop on an acre where corn could be grown carries risks that may include a new cropping system; no harvest, transport, or storage infrastructure; or no commodity market to fall back on if the biofuel market fails.

Estimated Production Costs

Production costs vary widely depending on tillage, irrigation, yield goal (soil fertility), spraying schedule, seed selection, and rotation. A sample corn budget with rain-fed, no-till, biotech seed, corn/soybean rotation, and 120 bushel yield goal would include a total cost of $211 per acre. If overhead – crop insurance, land, taxes – is included, the total is $305 per acre. Production costs increase to over $600 on irrigated fields with continuous corn.

Environmental and Sustainability Issues

Life cycle analysis (LCA) of ethanol production from corn grain has yielded a net energy ratio of 1.2 to 1.45, which represents just a 20% to 45% positive energy balance in producing ethanol from corn. A major criticism of corn ethanol has been the large amount of fossil energy used in production.

Environmental issues in corn production revolve around erosion, pesticide use, and nutrient use. Pesticides and nutrients have the potential to contaminate surface and ground water. Soil erosion has led to loss of topsoil and polluted streams and river systems with silt. Continuous attention to these issues has led to improvements, yet they will remain concerns in crop production.

Sugar Cane

Sugarcane is one of the most promising agricultural sources of biomass energy in the world. It is the most appropriate agricultural energy crop in most sugarcane producing countries due to its resistance to cyclonic winds, drought, pests and diseases, and its geographically widespread cultivation. Due to its high energy-to-volume ratio, it is considered one of nature's most effective storage devices for solar energy and the most economically significant energy crop. The climatic and physiological factors that limit its cultivation to tropical and sub-tropical regions have resulted in its concentration in developing countries, and this, in turn, gives these countries a particular role in the world's transition to sustainable use of natural resources.

Sugarcane is a highly efficient converter of solar energy, and has the highest energy-to-volume ratio among energy crops. Indeed, it gives the highest annual yield of biomass of all species. Roughly, 1 ton of Sugarcane biomass-based on Bagasse, foliage and ethanol output – has an energy content equivalent to one barrel of crude oil. Sugarcane produces mainly two types of biomass, Cane Trash and Bagasse. Cane Trash is the field residue remaining after harvesting the Cane stalk and Bagasse is the milling by-product which remains after extracting sugar from the stalk. The potential energy value of these residues has traditionally been ignored by policy-makers and masses in developing countries. However, with rising fossil fuel prices and dwindling firewood supplies, this material is increasingly viewed as a valuable renewable energy resource.

Sugar mills have been using Bagasse to generate steam and electricity for internal plant requirements while Cane Trash remains underutilized to a great extent. Cane Trash and Bagasse are produced during the harvesting and milling process of Sugarcane which normally lasts 6 to 7 months.

Around the world, a portion of the Cane Trash is collected for sale to feed mills, while freshly cut green tops are sometimes collected for farm animals. In most cases, however, the residues are burned or left in the fields to decompose. Cane Trash, consisting of Sugarcane tops and leaves can potentially be converted into around 1kWh/kg, but is mostly burned in the field due to its bulkiness and its related high cost for collection/ transportation.

On the other hand, Bagasse has been traditionally used as a fuel in the Sugar mill itself, to produce steam for the process and electricity for its own use. In general, for every ton of Sugarcane processed in the mill, around 190 kg Bagasse is produced. Low pressure boilers and low efficiency steam turbines are commonly used in developing countries. It would be a good business proposition to upgrade the present cogeneration systems to highly efficient, high pressure systems with higher capacities to ensure utilization of surplus Bagasse.

Sugarcane Trash as Biomass Resource

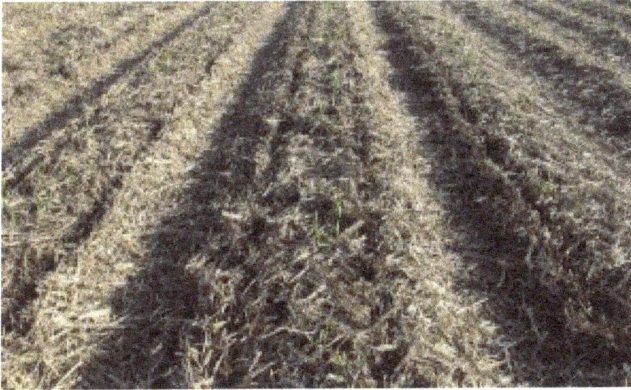

Sugarcane trash (or cane trash) is an excellent biomass resource in sugar-producing countries worldwide. The amount of cane trash produced depends on the plant variety, age of the crop at harvest and soil and weather conditions. Typically it represents about 15% of the total above ground biomass at harvest which is equivalent to about 10-15 tons per hectare of dry matter. During the harvesting operation around 70-80% of the cane trash is left in the field with 20-30% taken to the mill together with the sugarcane stalks as extraneous matter.

Cane trash's calorific value is similar to that of bagasse but has an advantage of having lower moisture content, and hence dries more quickly. Nowadays only a small quantity of this biomass is used as fuel, mixed with bagasse or by itself, at the sugar mill. The rest is burned in the vicinity of the dry cleaning installation, creating a pollution problem in sugar-producing nations.

Cane trash and bagasse are produced during the harvesting and milling process of sugarcane which normally lasts between 6 to 7 months. Cane trash can potentially be converted into heat and electrical energy. However, most of the trash is burned in the field due to its bulky nature and high cost incurred in collection and transportation.

Cane trash could be used as an off-season fuel for year-round power generation at sugar mills. There is also a high demand for biomass as a boiler fuel during the sugar-milling season. Sugarcane trash can also converted in biomass pellets and used in dedicated biomass power stations or co-firedwith coal in power plants and cement kilns.

Burning of cane trash creates pollution in sugar-producing countries

Currently, a significant percentage of energy used for boilers in sugarcane processing is provided by imported bunker oil. Overall, the economic, environmental, and social implications of utilizing cane trash in the final crop year as a substitute for bunker oil appears promising. It represents an opportunity for developing biomass energy use in the Sugarcane industry as well as for industries / communities in the vicinity.

Positive socio-economic impacts include the provision of large-scale rural employment and the minimization of oil imports. It can also develop the expertise necessary to create a reliable biomass supply for year-round power generation.

Recovery of Cane Trash

Recovery of cane trash implies a change from traditional harvesting methods; which normally consists of destroying the trash by setting huge areas of sugarcane fields ablaze prior to the harvest. There are a number of major technical and economic issues that need to be overcome to utilize cane trash as a renewable energy resource. For example, its recovery from the field and transportation to the mill, are major issues.

Alternatives include the current situation where the cane is separated from the trash by the harvester and the two are transported to the mill separately, to the harvesting of the whole crop with separation of the cane and the trash carried out at the mill. Where the trash is collected from the field it maybe baled incurring a range of costs associated with bale handling, transportation and storage. Baling also leaves about 10-20% (1-2 tons per hectare) of the recoverable trash in the field.

A second alternative is for the cane trash to be shredded and collected separately from the cane during the harvesting process. The development of such a harvester-mounted cane trash shredder and collection system has been achieved but the economics of this approach require evaluation. A third alternative is to harvest the sugarcane crop completely which would require an adequate collection, transport and storage system in addition to a mill based cleaning plant to separate the cane from the trash.

A widespread method for cane trash recovery is to cut the cane, chop into pieces and then it is blown in two stages in the harvester to remove the trash. The amount of trash that goes along with the cane is a function of the cleaning efficiency of the harvester. The blowers are adjusted to get adequate cleaning with a bearable cane loss.

On the average 68 % of the trash is blown out of the harvester, and stays on the ground, and 32 % is taken to the mill together with the cane as extraneous matter. The technique used to recover the trash staying on the ground is baling. Several baling machines have been tested with small, large, round and square bales. Cane trash can be considered as a viable fuel supplementary to bagasse to permit year-round power generation in sugar mills.

Thus, recovery of cane trash in developing nations of Asia, Africa and Latin America implies a change from traditional harvesting methods, which normally consists of destroying the trash by setting huge areas of cane fields ablaze prior to the harvest. To recover the trash, a new so-called "green mechanical harvesting" scheme will have to be introduced. By recovering the trash in this manner, the production of local air pollutants, as well as greenhouse gases contributing to adverse climatic change, from the fires are avoided and cane trash could be used as a means of regional sustainable development.

Soybeans

Soybean oil is a major feedstock for biodiesel production. Soybeans contain approximately 18% oil.

Soybean as a Biofuel Feedstock

Soybean (Glycine max) is a major crop throughout much of North America, South America and Asia. The United States is the world's greatest producer, producing approximately 32% of the worlds soybeans followed by Brazil with 28%. Origins of soybean are in Southeast Asia with first domestication reported in the 11th century BC in

China. First planted in the U.S. in 1765, soybeans spread to the Corn Belt by the mid 1800s with major acreage not seen until the 1920s when it was used mainly as a forage crop. Major U.S. expansion as an oilseed crop began in the 1940s. Soybeans contain approximately 18 to 20% oil compared to other oilseed crops such as canola (40%) and sunflower (43%). At 48 lbs per bushel soybean meal remains a major product from soybeans used for animal feed and human food. Soybean acreage is much greater than other oilseed crops leading to substantial soybean oil production and its availability as a biofuel feedstock.

Current Potential for Use as a Biofuel

Soybean oil is currently a major feedstock for production of biodiesel (NBB). The most common method of biodiesel production is a reaction of vegetable oils or animal fats with methanol or ethanol in the presence of sodium hydroxide (which acts as a catalyst). The transesterification reaction yields methyl or ethyl esters (biodiesel) and a byproduct of glycerin. Note biodiesel is not straight vegetable oil burned in a diesel engine. Numerous studies between 1980 and 2000 have shown the use of straight vegetable oil including soybean oil to cause carbon deposits and shorten engine life. Biodiesel use in diesel engines does not have similar negative effects. Use of soybean oil for biodiesel was greatly influenced by promotion from U.S. soybean farmers through the United Soybean Board (USB) and subsequent creation of the National Biodiesel Board (NBB).

Biology and Adaptation

Soybean (Glycine max) is a cool season legume which can be grown from south to north throughout much of the eastern half of the U.S. Soybeans and other legumes have a unique relationship with a bacteria bradyrhizobium species, will colonize on soybean roots forming a nodule. The two species form a symbiotic relationship where the soybean plant provides nutrition and the bacteria fixes nitrogen from the air. This relationship reduces the need for supplemental nitrogen fertilizer in soybean production.

Soybeans flower in response to day length and temperature. Varieties grown in the United States are divided into 13 maturity groups from maturity group 000 being the earliest and adapted to northern regions of Minnesota and southern Canada, to maturity group X adapted to southern regions such as south Texas. The earlier varieties bloom when days are long and nights are short, while the later-maturing varieties bloom under relatively shorter days and longer nights. Summer days are longer at northern latitudes, where early maturing varieties will initiate flowering when days are longer. maturity groups develop differently and knowing the growth habit of different maturity groups can help with the crop management.

Production and Agronomic Information

Through much of the upper Midwest soybeans are planted in April to June and harvested in September to November. Soybeans are well adapted to grow in soils similar to corn production. In many cases soybeans are grown in rotation with corn or wheat to break insect, weed, and disease cycles. Nutrient requirements are generally less for soybeans than other crops with major nutrient requirements nitrogen, phosphorous, and potassium and where much of the nitrogen is gained through a relationship with bacteria. A soil pH in the range of 5.5 to 7.0 will enhance nutrient availability and soybean growth. Weed control is necessary to achieve optimal yields and use of biotech seeds has eased the ability to control weeds during the growing season. Currently in the U.S. over 90% of soybeans planted are herbicide resistant. Many insects and diseases are common in soybeans grown in the upper Midwest. The most damaging pest to soybeans is soybean cyst nematode, a soil born parasitic roundworm feeding on soybean roots. Insect pests include: bean leaf beetle, soybean aphid, green clover worm, and spider mites. Soybean harvest begins after 95% leaf senescence when beans are 12 to 18% moisture.

Potential Biofuel Yields

Current U.S. production of soybeans in 2009 was 3.4 billion bushels from 77.4 million acres. Average yield per acre for the U.S. was 44 bushels per acre (National Ag Statistics Service). One bushel of soybeans can yield 1.5 gallons of biodiesel (NBB). Using all U.S. soybeans for biodiesel could produce 5.1 billion gallons biofuel. Using all soybean production for biodiesel has not been proposed and is not realistic. In 2009 biodiesel production was 700 million gallons with a production capacity of 1.83 billion gallons. Based on a yield of 44 bushels per acre, an acre of soybeans could yield 66 gallons of biodiesel compared to 69 gallons for a 1300 lb per acre canola yield, 84 gallons for sunflower and over 600 gallons for palm oil.

Production Challenges

Soybean production generally compliments corn production in the upper Midwest. Both corn and soybeans enjoy a long history of production on millions of acres in the upper Midwest. This history has led to a large infrastructure of equipment, storage, rail, barge and truck transportation. Soybeans like many crops face insect and disease pests along with weather related challenges. An emerging disease has gained much attention in recent years. Soybean rust, a fungal disease native to Asia has spread to the soybean fields of South America and finally to U.S. soybeans. Rust control is expensive requiring fungicide applications and yield damage can be extreme.

Estimated Production Costs

Production costs will vary depending on location, cropping systems, and fluctuation in price of energy. Major expenditures in soybean production include; Planting, harvesting, seed and pesticides. An example of a Nebraska rainfed budget for no-till soybeans for 2010 lists $115 per acre for field operations, materials and services. When including

overhead costs for land, insurance, etc. the total is approximately $200 per acre. To-
tal costs for irrigated soybeans are around $400 per acre. Biodiesel profitability is ex-
tremely variable and based on the continuously changing prices biodiesel, soybean oil,
co-product glycerin, methanol and natural gas. Price of soybean oil feedstock is one of
the driving factors in profitability.

Environmental and Sustainability Issues

Soybeans use for biodiesel production capacity grew from zero to over a billion gallons
per year in the past two decades 90s and 00s. Through that time biodiesel production
rose and fell depending on the price of feedstock, price of petroleum oil, and federal
and state subsidies provided to the industry. One major challenge for soybeans are
competing uses for soybean oil. Soybean oil is used for many human food products,
cooking oil and numerous industrial applications. Soybeans account for 80 percent or
more of the edible fats and oils consumed in the U.S. Competition with other uses has
caused price spikes in the soybean oil market challenging the profitability of soybean
biodiesel. A life cycle analysis of biodiesel done by the USDA reports the fossil energy
ratio of biodiesel to be 3.2 units of energy out for each unit of fossil energy used. In other words, biodiesel yields 3.2 units of energy for every unit of fossil energy consumed
compared to petroleum diesel which has a fossil energy ratio of ~0.84.

Vegetable Oil

Vegetable oil can be used to fuel diesel engines as straight vegetable oil (SVO) or as biodiesel following conversion. It can be used either for cars, cooking or for stationary use
to power machinery or an electricity generator. The more advanced the engine technology, the more complicated or limited the use of SVO. Hence for use in modern cars with
direct injection engines the conversion of SVO to biodiesel is mandatory. However SVO
can be directly used as fuel for most simple stationary diesel engines used in generator
sets or pump stations. Special SVO engines are also available. For robust stationary
(non-vehicle) engines, the use of pure SVO is the most attractive option. This way the
additional and somewhat dangerous biodiesel conversion process (due to some highly
inflammable components) can be avoided; furthermore the energy yield per gallon of
vegetable oil is higher when using SVO directly.

Listed below are the potential benefits of SVOs as fuel for power generation in rural
areas:

- SVOs are a renewable energy source low in greenhouse gas (GHG) emissions.

- Local production is possible and can contribute to value generation in rural
  areas.

- SVOs are liquid, hardly evaporate and are thus easily handled, stored and transported.

- SVOs are neither flammable nor explosive and do not release toxic gases.

- SVOs can be burnt in a relatively clean manner.

- SVOs do not cause major damage if accidentally spilt.

Almost all vegetable oils have a calorific value that is very similar to that of diesel fuel. However vegetable oil differs from diesel considerably with regard to viscosity, which is 10–20 times higher for SVOs. Furthermore, the solidification points of SVOs can be rather high. Coconut oil for example solidifies at temperatures below 22°C. This can cause major prob-ems particularly in cold regions. Oil characteristics can differ considerably between different types of SVOs as their composition of fatty acids varies. This can lead to plugging and gumming of lines, filters and injectors, as well as cause deposits inside the motor or excessive engine wear. There is long-term experience particularly with rapeseed oil, which has been used in Germany and other countries as fuel for cars, trucks and tractor engines for several years. There is no doubt that rapeseed oil is relatively easy and reliable to use in diesel engines. Other oils like palm oil, sunflower, soybean, coconut, jatropha and cotton oil have also been used to fuel motors. Many examples are regularly presented at conferences and can be found in publications. However there does not yet seem to be much large-scale and long-term practical experience regarding the proper and economic long-term functioning of such plant oil motor-generator systems. For economic long-term use all machines need clean homogenous oil with specific characteristics.

Sources of Vegetable Oil

Man currently uses over one thousand different oil plants. Some of their seeds are extracted from natural forests; some are cultivated in small plots or hedges, while others can be pro-duced in plantations with intensive agricultural methods. The most cultivated oil plants of the tropics are oil palm, coconut, soy, castor, sunflower and cotton. In many cases oil is only a by-product that is rarely extracted.

In the discussion on biofuels jatropha received particular focus due to its attributes:

- Wide adaptability to diverse climatic zones and soil types combined with high drought tolerance.

- Cultivable in areas where food production is not possible, therefore not competing with food crops.

- Short gestation period, easy multiplication and high pest and disease resistance.

The oil from cottonseed can be used as an energy source. Cotton oil can also be used for human consumption. However it is of minor value and due to the intensive use of chemical pest control in cotton culture its agrochemical content can be problematic for alimentation. In cases where cotton oil is available at reasonable prices it might serve as a sound source of energy. Usually cotton oil is not extracted as the cottonseed is used as forage for domestic animals.

Palm oil is produced at both industrial and small-scale levels. The oil palm is one of the most productive oil plants in terms of produced oil per hectare. Coconut and sunflower are also important sources of vegetable oil. All these oils have high value for human consumption and are therefore preferably used for nutritional purposes. However, in some particular circum-stances the energetic use of these oils might prove reasonable.

There are considerable advantages for a farmer if he has several options to market his oil produce. So in cases of low quality oil production or very low prices for edible oil he can sell or use it for energy purposes.

There are many other agricultural plants cultivated for food, fodder or fibre that con-tain oil. Some wild or semi-wild growing trees are also occasionally used as oil sources. In some cases this oil can be available at costs and in quantities that allow its use as fuel. The use of kapok oil in Cambodia is an example.

Besides fresh vegetable oil, waste oils - for example from restaurants - can also be used as a fuel source. Almost all countries (should) have regulatory frameworks that inhib-it irregular disposal of oils and fats. Hence the recycled fraction of these materials is high. Much of this is processed at industrial level. But examples of small-scale oil and biodiesel productions do exist.

Extraction

The two main processes of extracting oil from seed feedstock are mechanical press ex-trac-tion and solvent extraction.

The solvent process extracts more of the oil contained in the oilseed feedstock but re-quires costlier equipment. Oil extraction by a mechanical press is the standard tech-nology for small-scale applications. During mechanical press extraction, the oilseed is crushed and pressed in a screw press. The oilseed is heated either before or during the pressing. After most of the oil is removed, the remaining seed meal can be used as animal feed.

Hand presses are not appropriate for extraction of larger seed quantities. Screw press-es driven by combustion engines are standard and widely available in the market. The efficiency and durability of the different available presses will probably be a topic for future research and development.

After the crushing and pressing, the oil has to be filtered. Chamber filters with fine mesh tex tile filters are the most common option.

Market Development

The developments in the world biofuel market are important for any local approach. Many jatropha and other oil plant plantations have been established due to the biofuel hype of the recent years. Even though their yields do not fulfil the original expectations, these plantations will produce considerable amounts of oilseeds within the next few years. This will probably lead to an increasing availability of SVO on the markets; SVO production know-how will grow considerably with the lessons learnt from cultivation during this time.

And the gold rush mood is not yet over. New crops and plants are being touted as 'new oil wells,' e.g. the Acrocomia palm tree in Southern America was recently promoted as an in-sider tip. The implications of this gold rush mood are not yet clearly visible.

The demand for vegetable oils as a base for car fuels will persist and increase on the global market. There are some sectors like the aviation business with high purchasing power and technical applications that have no alternative to liquid fuels. Their demand for liquid fuels will therefore remain or increase further.

The development of environmental and social standards will most likely ensure that in future only certified SVO or biodiesel will find its way into the formal market. This will probably lead to high competition and low availability of cheap SVOs in the local markets of developing countries. However, some quantities of SVOs that do not meet international standards might remain for local use.

In any case the development of the global markets will increase infrastructure and know-how about SVO production and use, including high yield varieties, agricultural extension service, availability of appropriate engines etc. Local initiatives will probably also benefit from this development.

Cellulose

Cellulose is a polymer or string of glucose molecules joined together in linear rows. The parallel rows of glucose molecules form a tough crystalline substance which gives wood, paper and cardboard their strength. Cellulose comprises about 40-60% of the material in common forms of biomass, such as wood, paper, switchgrass, and corn stover. Cellulose is probably the most abundant organic molecule on the planet.

The other biomass components are hemicellulose, made from 5-carbon sugars like xylose, and lignin, which is a set of non-sugar molecules acting like a glue to hold the biomass molecules together. The 5-carbon sugars, or pentoses, can also be fermented to ethanol, and the lignin can be burned for energy or spun into high-strength carbon fibers for the light carbon composites used in wind turbine blades and aircraft parts With further cost reductions, carbon fibers made from biomass lignin should become part of an increasing new market for light weight automobiles, including EVs. Glucose, xylose and lignin from biomass are currently the only feasible alternatives for large-scale substitution of petroleum hydrocarbons.

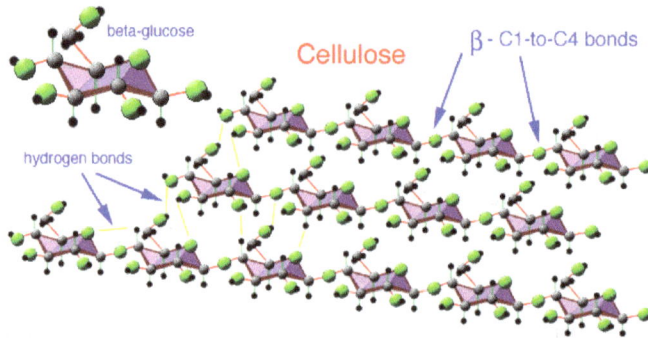

Cellulosic ethanol is fuel ethanol made from glucose, a 6-carbon sugar derived from the cellulose in biomass. Cellulosic ethanol is a substitute for gasoline. It is chemically identical to ethanol made from food crops like corn and sugar, but comes from wood, waste paper, and energy crops like poplar and switchgrass. Cellulosic ethanol is more difficult to make, because cellulose is a tough structural material, unlike starch from grains which is easily broken down to glucose sugar.

The primary feedstocks for fuel ethanol today include glucose from corn starch in the U.S., currently producing about 12 billion gallons of corn ethanol per year. Additional corn ethanol plants could push the total to 15 billion gal/year, with the downside of increasing corn prices and encouraging the shift of wheat and soybean acreage to corn. One bushel of corn produces about 2.8 gal ethanol, so 12 billion gal corn ethanol requires about 4.3 billion bu/year, a significant fraction of the U.S. corn crop.

In Brazil, the primary feedstock for fuel ethanol is sucrose (white sugar) from sugar cane, a more efficient source than corn. Standard gasoline in Brazil is about 25% ethanol, and flexfuel vehicles using E85 are popular. Growing large amounts of sugar cane is not an option in the U.S., Canada and Europe because of the tropical climate required.

Cellulosic ethanol can be made from a variety of biomass feedstocks including recycled paper, urban waste paper diverted from municipal solid waste (MSW), agricultural wastes like sugarcane bagasse and corn stover, energy crops like poplar, willow, and switchgrass, wood waste, and waste streams from pulp and paper mills. Cellulose and hemicellulose are the two carbohydrate components in most biomass, and can be hydrolyzed (split) into their component sugars: glucose, xylose, both of which can be fermented to ethanol, other biofuels like butanol, green diesel or biojetfuel, or lactic acid which can be made into biodegradable or more permanent plastics and fabrics. Cellulosic ethanol is also a sustainable feedstock for drop-in bioplastics like renewable plant-based polyethylene and polypropylene.

Cellulosic ethanol made from biomass will produce smaller amounts of greenhouse gases than corn ethanol, and far less than the gasoline it will replace. This greenhouse reduction potential can be estimated by Life Cycle Analysis (LCA), used by numerous groups to evaluate the net energy balance and carbon balance of ethanol from corn and cellulosic biomass.

The technologies can be broadly categorized in two path ways:

1. Biochemical conversion (fermentation) through pretreatment and hydrolysis and

2. Thermo-chemical conversion through gasification.

Biochemical Conversion

The first step is to break down cellulose which requires pretreatment. In pretreatment process, hemicelluloses and lignin that surround cellulose are broken down under a moderately high-temperature, high-pressure and through use of dilute acid. This process which is called hydrolysis breaks down hemicelluloses and dissolves lignin. The lignin thus produced forms and important source of heat and electricity, hence, limiting the use of fossil fuel in the conversion process. However, the problem with lignin is that it can under certain conditions during the pretreatment, redeposit onto cellulose which ultimately reduces the yield of sugar. Also this dilute acid treatment makes the process expensive as it requires costly equipment.

It has been proven that a milder pretreatment process whereby an appropriate mixture of enzyme to the hydrolysis of hemicellulose can curb degradation of sugar and ultimately the cost of processing.

Another way of enhancing the pretreatment process has been identified as the Ammonia Fiber Explosion (AFEX) process in which the lignocellulosic biomass is treated with high-pressure liquid ammonia leading to the explosive release of the pressure and thereby rendering the lignocellulosic biomass more susceptible to the enzymatic hydrolysis.

The simple sugars broken down from the cellulosic materials are fermented using yeast or bacteria under ideal conditions. These microorganisms convert the sugar into ethanol and water which is called the ethanol recovery process. The water is removed through distillation. The process is similar to that of bioethanol.

Thermo-chemical Conversion

In thermo-chemical conversion of cellulose into ethanol, the ligno-cellulosic raw material is broken into syngas i.e., carbon monoxide and hydrogen first applying heat and chemicals. This process is mostly appropriate when forest products and mill residues that are rich in lignin are used as feedstock and which cannot be converted by biochemical process. This process is however, complex and is similar to that of petrol refining in which contaminants (tar, sulphur, etc.) are also produced along with the syngas. The syngas are converted into ethanol which then undergoes distillation.

Feasibility of Technology and Operational Necessities

Production of cellulosic ethanol for large scale commercial use also requires cultivation of energy crops besides collection of waste as feedstocks. Therefore this requires adequate land, suitable soil and adequate water. Recent studies suggest of world's 13.5 billion hectares of surface lands, approximately 1.6 billion hectares of land is used as cropland. Therefore, availability of land besides forest land, protected areas and cultivated food cropland that could be used for producing energy crops is estimated between 250-800 million hectares and most of which lies in the tropical regions of Latin America and Africa.

The very first logistics for producing ethanol from cellulose requires producing biomass which can be obtained from forest resources and/or agricultural resources. The produced biomass has to be harvested and then stored before it is finally transported to the conversion plant. These steps comprise theupstream logistics.

Before the biomass is introduced to the bio-refinery, the size in which they've been harvested needs to be reduced to the level where it is easy to handle and the process becomes more efficient. So if it is agricultural residue they need to be grinded and if wooden residue then these has to be taken through the chipping process so that the size of the materials are uniform.

Some of the key barriers to the commercialization of cellulosic ethanol as as under:

- Infrastructure barriers: Infrastructure for the distribution of cellulosic ethanol might be a barrier because it cannot be transported through pipes but has to be carried through vehicles by road. Georgraphical locations suitable for production of feedstock and the area of demand for the ethanol may vary thus requiring extensive investment in infrastructure development like road/railway networks. Also greater investment required for storing and processing especially in the pre-treatment stage.

- Market barriers: Lack of distribution infrastructure for market penetration, availability of a limited number of vehicle models that can run with the given type of fuel combination that uses high ethanol blend percentage and lack of awareness about the benefits of the cellulosic ethanol among the industry participants as well as the consumers are barriers related with the market place.

- Feedstock logistical and technical barriers: At present there is lack of efficiency in meeting the requirements related with sustainable harvesting of the feedstock and development of bio-refineries, harvest equipment and technology which can deal with the large yield of biomass. Production of cellulosic ethanol requires adequate land, water and suitable soil.

- Financial barriers: Expensive process and requires costly equipment, a high cost of cellulose enzyme is a barrier for economic production.

Algal Oil

Generally, algae are a diverse group of prokaryotic and eukaryotic organisms ranging from unicellular genera such as Chlorella and diatoms to multicellular forms such as the giant kelp, a large brown alga that may grow up to 50 m in length. Algae can either be autotrophic or heterotrophic. The autotrophic algae require only inorganic compounds such as CO_2, salts, and a light energy source for their growth, while the heterotrophs are non-photosynthetic, which require an external source of organic compounds as well as nutrients as energy sources. Microalgae are very small in sizes usually measured in micrometers, which normally grow in water bodies or ponds. Microalgae contain more lipids than macroalgae and have the faster growth in nature.

The algae can be converted into various types of renewable biofuels including bioethanol, biodiesel, biogas, photobiologically produced biohydrogen, and further processing for bio-oil and syngas production through liquefaction and gasification, respectively. The conversion technologies for utilizing algal biomass to energy sources can be categorized into three different ways, i.e., biochemical, chemical, and thermochemical conversion and make an algal biorefinery, which has been depicted in figure below.

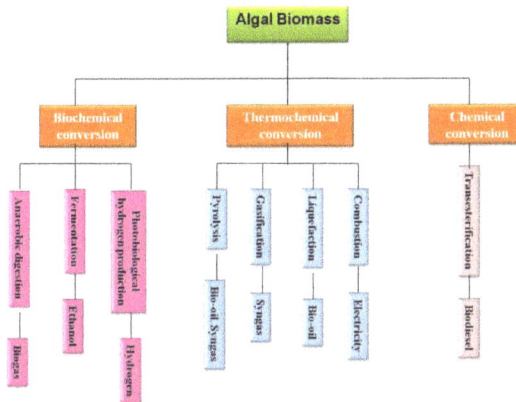

Algal biomass conversion process for biofuel production

Algal biodiesel production involves biomass harvesting, drying, oil extraction, and further transesterification of oil, which have been described as below.

Harvesting and Drying of Algal Biomass

Unicellular microalgae produce a cell wall containing lipids and fatty acids, which differ them from higher animals and plants. Harvesting of algal biomass and further drying is important prior to mechanical and solvent extraction for the recovery of oil. Macroalgae can be harvested using nets, which require less energy while microalgae can be harvested by some conventional processes, which include filtration flocculation, centrifugation, foam fractionation, sedimentation, froth floatation, and ultrasonic separation. Selection of harvesting method depends on the type of algal species.

Drying is an important method to extend shelf-life of algal biomass before storage, which avoids post-harvest spoilage. Most of the efficient drying methods like spray-drying, drum-drying, freeze drying or lyophilization, and sun-drying have been applied on microalgal biomass. Sun-drying is not considered as a very effective method due to presence of high water content in the biomass. However, Prakash et al. used simple solar drying device and succeed in drying Spirulina and Scenedesmus having 90% of moisture content. Widjaja et al. showed the effectiveness of drying temperature during lipid extraction of algal biomass, which affects both concentration of triglycerides and lipid yield. Further, all these processes possess safety and health issues.

Extraction of Oil from Algal Biomass

Unicellular microalgae produce a cell wall containing lipids and fatty acids, which differ them from higher animals and plants. In the literature, there are different methods of oil extraction from algae, such as mechanical and solvent extraction. However, the extraction of lipids from microalgae is costly and energy intensive process.

Mechanical oil extraction. The oil from nuts and seeds is extracted mechanically using presses or expellers, which can also be used for microalgae. The algal biomass should be dried prior to this process. The cells are just broken down with a press to leach out the oil. About 75% of oil can be recovered through this method and no special skill is required. Topare et al. extracted oil through screw expeller by mechanical pressing (by piston) and osmotic shock method and recovered about 75% of oil from the algae. However, more extraction time is required as compared to other methods, which make the process unfavorable and less effective.

Solvent based oil extraction. Oil extraction using solvent usually recovers almost all the oil leaving only 0.5–0.7% residual oil in the biomass. Therefore, the solvent extraction method has been found to be suitable method rather than the mechanical extraction of oil and fats. Solvent extraction is another method of lipid extraction from microalgae, which involves two stage solvent extraction systems. The amount of lipid extracted from microalgal biomass and further yield of highest biodiesel depends mainly on the solvent used. Several organic solvents such as chloroform, hexane, cyclo-hexane, acetone, and benzene are used either solely or in mixed form. The solvent reacts on algal cells releasing oil, which is recovered from the aqueous medium. This occurs due to the nature of higher solubility of oil in organic solvents rather than water. Further, the oil can be separated from the solvent extract. The solvent can be recycled for next extraction. Out of different organic solvents, hexane is found to be most effective due to its low toxicity and cost.

In case of using mixed solvents for oil extraction, a known quantity of the solvent mixture is used, for example, chloroform/methanol in the ratio 2:1 (v/v) for 20 min using a shaker and followed by the addition of mixture, i.c., chloroform/water in the ratio of 1:1 (v/v) for 10 min. Similarly, Pratoomyot et al extracted oil from different algal species using the solvent system chloroform/methanol in the ratio of 2:1 (v/v) and found different fatty acid content. Ryckebosch et al. optimized an analytical procedure and found chloroform/methanol in the ratio 1:1 as the best solvent mixture for the extraction of total lipids. Similarly, Lee et al. extracted lipid from the green alga Botryococcus braunii using different solvent system and obtained the maximum lipid content with chloroform/methanol in the ratio of 2:1. Hossain et al., used hexane/ether in the ratio 1:1 (v/v) for oil extraction and allowed to settle for 24 h. Using a two-step process, reported about 80% of lipid recovery using ethanol and hexane in the two steps for the extraction and purification of lipids. Therefore, a selection of a most suitable solvent system is required for the maximum extraction of oil for an economically viable process.

Lee compared the performance of various disruption methods, including autoclaving, bead-beating, microwaves, sonication, and using 10% NaCl solution in the extraction of Botryococcus sp., Chlorella vulgaris, and Scenedesmus sp, using a mixture of chloroform and methanol (1:1).

Transesterification

This is a process to convert algal oil to biodiesel, which involves multiple steps of reactions between triglycerides or fatty acids and alcohol. Different alcohols such as ethanol, butanol, methanol, propanol, and amyl alcohol can be used for this reaction. However, ethanol and methanol are used frequently for the commercial development due to its low cost and its physical and chemical advantages. The reaction can be performed in the presence of an inorganic catalyst (acids and alkalies) or lipase enzyme. In this method, about 3 mol of alcohol are required for each mole of triglyceride to produce 3 mol of methyl esters (biodiesel) and 1 mol of glycerol (by-product). Glycerol is denser than biodiesel and can be periodically or continuously removed from the reactor in order to drive the equilibrium reaction. The presence of methanol, the co-solvent that keeps glycerol and soap suspended in the oil, is known to cause engine failure. Thus, the biodiesel is recovered by repeated washing with water to remove glycerol and methanol.

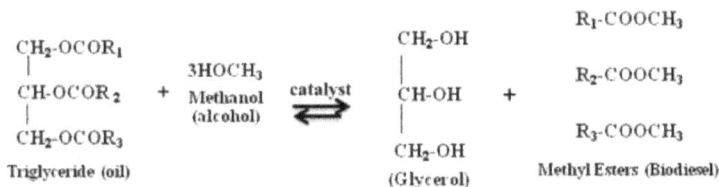

Figure: Transesterification of oil to biodiesel. R_{1-3} are hydrocarbon groups.

The reaction rate is very slow by using the acid catalysts for the conversion of triglycerides to methyl esters, whereas the alkali-catalyzed transesterification reaction has been reported to be 4000 times faster than the acid-catalyzed reaction. Sodium and potassium hydroxides are the two commercial alkali catalysts used at a concentration of about 1% of oil. However, sodium methoxide has become the better catalyst rather than sodium hydroxide.

used Scenedesmus sp. for the biodiesel production through acid and alkali transesterification process. They reported $55.07 \pm 2.18\%$, based on lipid by wt of biodiesel conversion using NaOH as an alkaline catalyst than using H_2SO_4 as $48.41 \pm 0.21\%$ of biodiesel production. In comparison to acid and alkalies, lipases as biocatalyst have different advantages as the catalysts due to its versatility, substrate selectivity, regioselectivity, enantioselectivity, and high catalytic activity at ambient temperature and pressure. It is not possible by some lipases to hydrolyze ester bonds at secondary positions, while some other group of enzymes hydrolyzes both primary and secondary esters. Another group of lipases exhibits fatty acids selectivity, and allow to cleave ester bonds at particular type of fatty acids. cloned the lipase gene lipB68 and expressed in Escherichia coli BL21 and further used it as a catalyst for biodiesel production. LipB68 could catalyze the transesterification reaction and produce biodiesel with a yield of 92% after 12 h, at a temperature of 20°C. The activity of the lipase enzyme with such a low temperature could provide substantial savings in energy consumption. However, it is rarely used due to its high cost.

Extractive transesterification. It involves several steps to produce biodiesel such as drying, cell disruption, oils extraction, transesterification, and biodiesel refining. The main problems are related with the high water content of the biomass (over 80%), which overall increases the cost of whole process.

In situ transesterification.This method skips the oil extraction step. The alcohol acts as an extraction solvent and an esterification reagent as well, which enhances the porosity of the cell membrane. Yields found are higher than via the conventional route, and waste is also reduced. Industrial biodiesel production involves release of extraction solvent, which contributes to the production of atmospheric smog and to global warming. Thus, simplification of the esterification processes can reduce the disadvantages of this attractive bio-based fuel. The single-step methods can be attractive solutions to reduce chemical and energy consumption in the overall biodiesel production process. A comparison of direct and extractive transesterification is given in table below.

Table: Comparison of extractive transesterification and *in situ* methods

Sl. no.	Extractive transesterification	In situ transesterification
1	Low heating value	Heating value is high
2	Product yield is low	Higher product yield
3	Process is complex and time taking	Quick and simple operation process
4	Lipid loss during process	Avoided potential lipid loss
5	Waste water pollutes the environment	Reduced waste water pollutants
6	Production cost is high	Absence of harvesting and dewatering lowers the cost

Bioethanol Production

Several researchers have been reported bioethanol production from certain species of algae, which produce high levels of carbohydrates as reserve polymers. Owing to the presence of low lignin and hemicelluloses content in algae in comparison to lignocellulosic biomass, the algal biomass have been considered more suitable for the bioethanol production. Recently, attempts have been made (for the bioethanol production) through the fermentation process using algae as the feedstocks to make it as an alternative to conventional crops such as corn and soyabean. A comparative study of algal biomass and terrestrial plants for the production of bioethanol has been given in table. There are different micro and macroalgae such as Chlorococcum sp., Prymnesium parvum, Gelidium amansii, Gracilaria sp., Laminaria sp., Sargassum sp., and Spirogyra sp., which have been used for the bioethanol production. These algae usually require light, nutrients, and carbon dioxide, to produce high levels of polysaccharides such as starch and cellulose. These polysaccharides can be extracted to fermentable sugars through hydrolysis and further fermentation to bioethanol and separated through distillation as shown in figure below.

Table: Comparative study between algal biomass and terrestrial plants for bioethanol production

Feedstock	Conditions	Bioethanol
ALGAE		
Chlorococcum infusionum	Alkaline pre-treatment, temp. 120°C, S. cerevisiae	260 g ethanol/kg algae
Spirogyra	Alkaline pre-treatment, synthetic media growth, saccharification of biomass by Aspergillus niger, fermentation by S. cerevisiae	80 g ethanol/kg algae
Chlorococcum humicola	Acid pre-treatment, temp. 160°C, S. cerevisiae	520 g ethanol/kg microalgae
TERRESTRIAL PLANTS		
Madhuca latifolia	Strain Zymomonas mobilis MTCC 92, immobilized in Luffa cylindrical sponge disks, temp. 30°C	251.1 ± 0.012 g ethanol/kg flowers
Manihot esculenta	Enzyme termamyl and amyloglucosidase, 1 N HCl, Saccharomyces cerevisiae, ca-alginate immobilization	189 ± 3.1 g ethanol/kg flour cassava
Sugarcane bagasse	Acid (H_2SO_4) hydrolysis, Kluyveromyces sp. IIPE453, Fermentation at 50°C	165 g ethanol/kg bagasse
Rice straw	Cellulase, β-glucosidase, solid state fermentation, strain Trichoderma reesei RUT C30, and Aspergillus niger MTCC 7956	93 g ethanol/kg pretreated rice straw

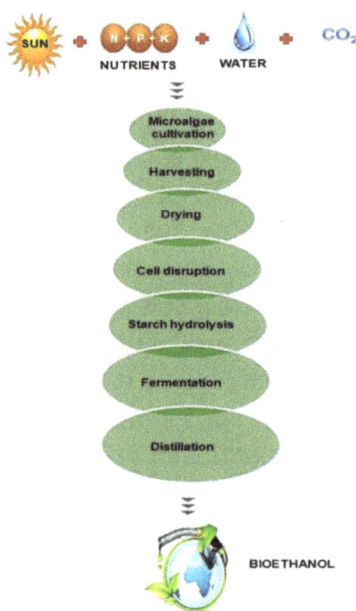

Process for bioethanol production from microalgae.

Pre-treatment and Saccharification

It has been reported that, the cell wall of some species of green algae like Spirogyra and Chlorococcumcontain high level of polysaccharides. Microalgae such as C. vulgaris contains about 37% of starch on dry weight basis, which is the best source for bioethanol with 65% conversion efficiency. Such polysaccharide based biomass requires additional processing like pre-treatment and saccharification before fermentation. Saccharification and fermentation can also be carried out simultaneously using an amylase enzyme producing strain for the production of ethanol in a single step. Bioethanol from microalgae can be produced through the process, which is similar to the first generation technologies involving corn based feedstocks. However, there is limited literature available on the fermentation process of microalgae biomass for the production of bioethanol.

The pre-treatment is an important process, which facilitates accessibility of biomass to enzymes to release the monosaccharides. Acid pre-treatment is widely used for the conversion of polymers present in the cell wall to simple forms. The energy consumption in the pre-treatment is very low and also it is an efficient process found 7% (w/w) H_2SO_4 as the promising concentration for the pre-treatment of the brown macroalgae Nizimuddinia zanardini to obtain high yield of sugars without formation of any inhibitors. Candra and studied the bioethanol production using red seaweed Eucheuma cottonii through acid hydrolysis. In this study, 5% H_2SO_4 concentration was used for 2h at 100°C, which yielded 15.8 g/L of sugars. However, there are other alternatives to chemical hydrolysis such as enzymatic digestion and gamma radiation to make it more sustainable.

Similar to starch, there are certain polymers such as alginate, mannitol, and fucoidan present in the cell wall of various algae, which requires additional processing like pre-treatment and saccharification before fermentation. Another form of storage carbohydrate found in various brown seaweeds and microalgae is laminarin, which can be hydrolyzed by β-1,3-glucanases or laminarinases. Laminarinases can be categorized into two groups such as exo- and endo-glucanases based on the mode of hydrolysis, which usually produces glucose and smaller oligosaccharides as the end product. Both the enzymes are necessary for the complete digestion of laminarin polymer.

Markou et saccharified the biomass of Spirulina (Arthrospira platensis), fermented the hydrolyzate and obtained the maximum ethanol yield of 16.32 and 16.27% ($g_{ethanol}/g_{biomass}$) produced after pre-treatment with 0.5 N HNO_3 and H_2SO_4, respectively. Yanagisawa et al. investigated the content of polysaccharide materials present in three types of seaweeds such as sea lettuce (Ulva pertusa), chigaiso (Alaria crassifolia), and agar weed (Gelidium elegans). These seaweeds contain no lignin, which is a positive signal for the hydrolysis of polysaccharides without any pre-treatment. Singh and used Spirogyra biomass for the production of bioethanol using Saccharomyces cerevisiae and Zymomonas mobilis. In a method, they followed acid pre-treatment of algal biomass and further saccharified using α-amylase producing Aspergillus niger. In another method, they directly saccharified the biomass without any pre-treatment. The direct saccharification process resulted in 2% (w/w) more alcohol in comparison to pretreated and saccharified algal biomass. This study revealed that the pre-treatment with different chemicals are not required in case of Spyrogyra, which reflects its economic importance for the production of ethanol. Also, cellulase enzyme has been used for the saccharification of algal biomass containing cellulose. However, this enzyme system is more expensive than amylases and glucoamylases, and doses required for effective cellulose saccharification are usually very high. applied different cellulases on green alga Ulva for saccharification and found highest conversion efficiency of biomass into reducing sugars by using cellulase 22119 rather than viscozyme L, cellulase 22086 and 22128.

Fermentation

There are different groups of microorganisms like yeast, bacteria, and fungi, which can be used for the fermentation of the pretreated and saccharified algal biomass under anaerobic process for the production of bioethanol. Nowadays, S. cerevisiae and Z. mobilis have been considered as the bioethanol fermenting microorganisms. However, fermentation of mannitol, a polymer present in certain algae is not possible in anaerobic condition using these well known microorganisms and requires supply of oxygen during fermentation, which is possible only by Zymobacter palmae.

Certain marine red algae contain agar, a polymer of galactose and galactopyranose, which can be used for the production of bioethanol. The biomass of red algae can be depolymerized into different monomeric sugars like glucose and galactose. In addition to mannitol and glucose, brown seaweeds contain about 14% of extra carbohydrates in the form of alginate. Horn et al. reported the presence of alginate, laminaran, mannitol, fucoidan, and cellulose in some brown seaweeds, which are good source of sugars. They fermented brown seaweed extract having mannitol using bacteria Z. palmae and obtained an ethanol yield of about 0.38 g ethanol/g mannitol.

In the literature, there are many advantages supporting microalgae as the promising substrate for bioethanol production. Hon-Nami used Chlamydomonas perigranulata algal culture and obtained different by-products such as ethanol and butanediol. Similarly, Yanagisawa et al. obtained glucose and galactose through the saccharification of agar weed (red seaweed) containing glucan and galactan and obtained 5.5% of ethanol concentration through fermentation using S. cerevisiae IAM 4178. Harun et al. obtained 60% more ethanol in case of lipid extracted microalgal biomass rather than intact algal biomass of Chlorococcum sp. This shows the importance of algal biomass for the production of both biodiesel and bioethanol.

Biogas Production

Recently, biogas production from algae through anaerobic digestion has received a remarkable attention due to the presence of high polysaccharides (agar, alginate, carrageenan, laminaran, and mannitol) with zero lignin and low cellulose content. Mostly, seaweeds are considered as the excellent feedstock for the production of biogas. Several workers have demonstrated the fermentation of various species of algae like Scenedesmus, Spirulina, Euglena, and Ulva for biogas production. The production of biogas using algal biomass in comparison to some terrestrial plants is shown in table below.

Table: Comparative study between algal biomass and terrestrial plants for biogas production

Feedstock	Conditions	Biogas
ALGAE		
Blue algae	pH-6.8, microcystin (MC) biodegradation	189.89 mL/g of VS
Chlamydomonas reinhardtii	Drying as the pre-treatment, batch fermentation, temp. 38°C	587 mL/g of VS
Scenedesmus obliquus		287 mL/g of VS
Ulva sp.	Batch reactor, Co-digestion with bovine slurry, temp. 35°C	191 mL/g of VS
Laminaria digitata		246 mL/g of VS
Saccorhiza polyschides		255 mL/g of VS

Feedstock	Conditions	Biogas
Saccharina latis-sima		235 mL/g of VS
TERRESTRIAL PLANTS		
Banana stem	Pre-treatment: 6% NaOH in 55°C for 54 h. 37 ± 1°C for 40 days, batch	357.9 mL/g of VS
Saline creeping wild ryegrass	35°C for 33 days, batch	251 mL/g of VS
Rice straw	Pre-treatment: ammonia conc. 4% and moisture content 70%, temp. 35 ± 2°C, 65 days, 120 rpm, batch	341.35 mL/g of VS
Date palm tree wastes	Pre-treatment: alkaline, particle size 2–5 mm, temp. 40°C	342.2 mL/g of VS

Biogas is produced through the anaerobic transformation of organic matter present in the biodegradable feedstock such as marine algae, which releases certain gases like methane, carbon dioxide, and traces of hydrogen sulfide. The anaerobic conversion process involves basically four main steps. In the first step, the insoluble organic material and higher molecular mass compounds such as lipids, carbohydrates, and proteins are hydrolyzed into soluble organic material with the help of enzyme released by some obligate anaerobes such as Clostridia and Streptococci. The second step is called as acidogenesis, which releases volatile fatty acids (VFAs) and alcohols through the conversion of soluble organics with the involvement of enzymes secreted by the acidogenic bacteria. Further, these VFAs and alcohols are converted into acetic acid and hydrogen using acetogenic bacteria through the process of acetogenesis, which finally metabolize to methane and carbon dioxide by the methanogens.

Sangeetha et al. reported the anaerobic digestion of green alga Chaetomorpha litorea with generation of 80.5 L of biogas/kg of dry biomass under 299 psi pressure. Vergara-Fernandez et al. evaluated digestion of the marine algae Macrocystis pyrifera and Durvillaea antarctica marine algae in a two-phase anaerobic digestion system and reported similar biogas productions of 180.4 (±1.5) mL/g dry algae/day with a methane concentration around 65%. However, in case of algae blend, same methane content was observed with low biogas yield. Mussgnug et al. reported biogas production from some selected green algal species like Chlamydomonas reinhardtii and Scenedesmus obliquus and obtained 587 and 287 mL biogas/g of volatile solids, respectively. Further, there are few studies, which have been conducted with microalgae showing the effect of different pre-treatment like thermal, ultrasound, and microwave for the high production of biogas.

However, there are different factors, which limit the biogas production such as requirement of larger land area, infrastructure, and heat for the digesters. The proteins present in algal cells increases the ammonium production resulting in low carbon to nitrogen ratio, which affects biogas production through the inhibition of growth of anaerobic microorganisms. Also, anaerobic microorganisms are inhibited by the sodium ions. Therefore, it is recommended to use the salt tolerating microorganisms for the anaerobic digestion of algal biomass.

Biohydrogen Production

Recently, algal biohydrogen production has been considered to be a common commodity to be used as the gaseous fuels or electricity generation. Biohydrogen can be produced through different processes like biophotolysis and photo fermentation. Biohydrogen production using algal biomass is comparative to that of terrestrial plants. Park et al. found Gelidium amansii (red alga) as the potential source of biomass for the production of biohydrogen through anaerobic fermentation. Nevertheless, they found 53.5 mL of H_2 from 1 g of dry algae with a hydrogen production rate of 0.518 L H_2/g VSS/day. The authors found an inhibitor, namely, 5-hydroxymethylfurfural (HMF) produced through the acid hydrolysis of G. amansii that decreases about 50% of hydrogen production due to the inhibition. Thus, optimization of the pre-treatment method is an important step to maximize biohydrogen production, which will be useful for the future direction. Saleem et al. reduced the lag time for hydrogen production using microalgae Chlamydomonas reinhardtii by the use of optical fiber as an internal light source. In this study, the maximum rate of hydrogen production in the presence of exogenic glucose and optical fiber was reported to be 6 mL/L culture/h, which is higher than other reported values.

Table: Comparative study between algal biomass and terrestrial plants for biohydrogen production.

Feedstock	Conditions	Biohydrogen
ALGAE		
Gelidium amansii	Hydrolysis at 150°C	53.5 mL of H_2/g of dry algae
Laminaria japonica	Mesophilic condition (35 ± 1°C), pH of 7.5, anaerobic sequencing batch reactor, hydraulic retention time (HRT) of 6 days	71.4 mL H_2/g of dry algae
TERRESTRIAL PLANTS		
Bagasse	Strain Klebsiella oxytoca HP1, temp. 37.5°C, pH-7	107.8 ± 7.5 mL H_2/g bagasse
Corn stalk	Temp. 55°C, pH-7.4	61.4 mL/g of cornstalk
Pretreated wheat straw	Strain Caldicellulosiruptor saccharolyticus, Temp. 70°C, pH-7.2	44.7 mL/g of dry wheat straw
Wheat straw	Acid pre-treatment, simultaneous saccharification and fermentation (SSF)	141 mL/g VS

Some of microalgae like blue green algae have glycogen instead of starch in their cells. This is an exception, which involves oxidation of ferrodoxin by the hydrogenase enzyme activity for the production of hydrogen in anaerobic condition. However, another function of this enzyme is to be involved in the detachment of electrons. Therefore, different researchers have focused for the identification of these enzyme activities having interactions with ferrodoxin and the other metabolic functions for microalgal photo-biohydrogen production. They are also involved with the change of these interactions genetically to enhance the biohydrogen production.

Bio-Oil and Syngas Production

Bio-oil is formed in the liquid phase from algal biomass in anaerobic condition at high temperature. The composition of bio-oil varies according to different feedstocks and processing conditions, which is called as pyrolysis. There are several parameters such as water, ash content, biomass composition, pyrolysis temperature, and vapor residence time, which affect the bio-oil productivity. However, due to the presence of water, oxygen content, unsaturated and phenolic moieties, crude bio-oil cannot be used as fuel. Therefore, certain treatments are required to improve its quality. Bio-oils can be processed for power generation with the help of external combustion through steam and organic rankine cycles, and stirling engines. However, power can also be generated through internal combustion using diesel and gas-turbine engines. In literature, there are limited studies on algae pyrolysis compared to lignocellulosic biomass. Although, high yields of bio-oil occur through fluidized-bed fast pyrolysis processes, there are several other pyrolysis modes, which have been introduced to overcome their inherent disadvantages of a high level of carrier gas flow and excessive energy inputs investigated suitability of the microalgal biomass for bio-oil production and found the superior quality than the wood. produced bio-oil from pyrolysis of algae (Nannochloropsis sp.) at 300°C after lipid extraction, which composed of 50 wt% acetone, 30 wt% methyl ethyl ketone, and 19 wt% aromatics such as pyrazine and pyrrole. Similarly, Choi et al. carried out pyrolysis study on a species of brown algae Saccharina japonica at a temperature of 450°C and obtained about 47% of bio-oil yield.

Gasification is usually performed at high temperatures (800–1000°C), which converts biomass into the combustible gas mixture through partial oxidation process, called syngas or producer gas. Syngas is a mixture of different gases like CO, CO_2, CH_4, H_2, and N_2, which can also be produced through normal gasification of woody biomass. In this process, biomass reacts with oxygen and water (steam) to generate syngas. It is a low calorific gas, which can be utilized in the gas turbines or used directly as fuel. Different variety of biomass feedstocks can be utilized for the production of energy through the gasification process.

Camelina

Camelina [Camellia sativa (L.) Crantz], also known as false flax or gold-of-pleasure, is a broadleaf oilseed flowering plant of the Brassicaceae (mustard) family that grows optimally in temperate climates. Other more common members of this family include important food crops such as broccoli, Brussels sprouts, cabbage, cauliflower, collards, kales, kohlrabi, radish, rapeseed/canola, rutabaga, turnip and various mustards. Camelina can be grown in a variety of climatic and soil conditions as a spring or summer annual or as a biannual winter crop.

Camelina Oil and Meal

Camelina yields anywhere from 336 to 2240 kg of seeds per hectare at maturity with the lipid content of individual seeds ranging between 35 and 45 weight percent (wt%). The resulting yield of camelina oil is thus calculated to be between 106 and 907 liters per hectare, which as seen in table below at the higher end of the range. It is higher than soybean and sunflower oils but lower than rapeseed oil. The wide variability in yield is attributable to such factors as the climate and conditions at the growing location as well as by the amount of agricultural inputs applied to the crop and the presence of pests, weeds and disease. With further genetic modification, the yield of camelina may approach that of rapeseed. Commercially, Great Plains Oil & Exploration, LLC and Sustainable Oils, LLC are the two primary commercial suppliers of camelina oil in the United States.

Table: Comparison of yields from several oilseeds

	Camelina	Rapeseed	Soybean	Sunflower
Seed yield (T/ha)	0.90–2.24	2.68–3.39a	2.14–2.84a	1.44–1.70a
Oil content (wt%)	35–45	40–44	18–22	39–49
Oil yield (L/ha)b	106–907	965–1342	347–562	505–750

    a   Oil World Annual 2009, 1, Oilseeds 5–9 (T = metric ton = 1000

          kg). Data from EU-27 from 2002–2006.

    b   Calculated by the author (BRM) from yield and oil content data.

α Linolenic acid generally comprises between 32 and 40 wt% of the fatty acid composition of camelina oil (Table). Other fatty acids in quantities above 10 wt% include linoleic, oleic, and 11-eicosenoic acids. The acid value of crude camelina oil extracted either by solvent (hexane) or cold press is typically between 1 and 5 mg KOH/g of oil, which is considerably lower than many other crude plant oils. Camelina oil has been used for numerous applications including directly as an experimental fuel for diesel transport engines, as a culinary oil, and as a biological feedstock for the experimental production of longchain wax esters for potential cosmetic and lubricant applications. Camelina oil may also be used as an industrial source of a-linolenic acid, as it has a comparatively high content of this constituent versus the commodity oils.

Extraction of oil from camelina seeds by mechanical expeller yields a meal that consists of approximately 10% residual oil, 45% crude protein, 13% fibers, 5% minerals, and other minor constituents such as glucosinolates and vitamins. The concentration of glucosinolates in dry camelina seeds ranges from 13 to 36 lmol/g. When ingested in sufficient quantity, glucosinolates cause deleterious effects in animals such as reduced palatability as well as decreased growth and production. Consequently, camelina meal cannot exceed 10 wt% of the total food ration given to feedlot beef cattle and broiler chickens in the United States. However, even a 10% ceiling represents significant market potential where large numbers of animals are raised for human consumption, thus representing an important revenue stream for camelina meal.

Table: Fatty acid composition (%) of camelina oil from several literature					
Fatty acid	Carbon number	Leonard	Moser & Vaughn	Zuhr & Matthaus	Frohlich & Rice
Palmitic	16:0	5.3	6.8	5.4	5.4
Stearic	18:0	2.5	2.7	2.5	2.6
Oleic	18:1	12.6	18.6	14.9	14.3
Linoleic	18:2	15.6	19.6	15.2	14.3
α-Linolenic	18:3	37.5	32.6	36.8	38.4
Arachidic	20:0	1.2	1.5	1.3	1.4
11-Eicosenoic	20:1	15.5	12.4	15.5	16.8
Eicosadienoic	20:2	2.0	1.3	1.9	
Eicosatrienoic	20:3	1.7	0.8	1.6	
Behenic	22:0	0.3	0.2	0.3	0.2
Erucic	22:1	2.9	2.3	2.8	2.9
Others	–	2.9	1.2	1.8	3.7

Camelina Oil as Feedstock for Biodiesel

Camelina is particularly attractive as an alternative feedstock for biodiesel production as a result of its low cost versus commodity oils coupled with its potential to significantly enhance domestic feedstock availability.

Biodiesel is classically prepared by transesterification of lipids in the presence of an homogenous alkali catalyst and excess methanol at elevated (608C) temperature [3–5]. figure below illustrates a typical schematic for industrial production of biodiesel. Crude feedstocks that have low free fatty acid contents (less than 3.0 wt%) can be directly transesterified without pretreatment, thereby eliminating a costly pretreatment step. Camelina is one such crude oil and has been successfully converted to biodiesel (fatty acid methyl esters) by the classic method as well as with heterogeneous metal oxide catalysts both with and without microwave irradiation and at non-catalytic sub- and supercritical conditions employing co-solvents with methanol. The fuel properties (cold flow properties, oxidative stability, kinematic viscosity, cetane number, etc) of camelina-based biodiesel are similar to those of biodiesel prepared from soybean oil, thus indicating its acceptability for use as biodiesel. Additionally, fatty acid ethyl esters have been prepared from camelina oil and along with methyl esters were evaluated as blend components in petrodiesel (a15 ppm S). As was the case with the neat esters, camelina-based biodiesel blends in petrodiesel exhibited fuel properties comparable to the corresponding soybean-based blends. Commercially, Green Earth Fuels, LLC in partnership with Sustainable Oils, LLC and INEOS in partnership with Great Plains Oil & Exploration, LLC are among the industrial producers of fatty acid methyl esters from camelina oil.

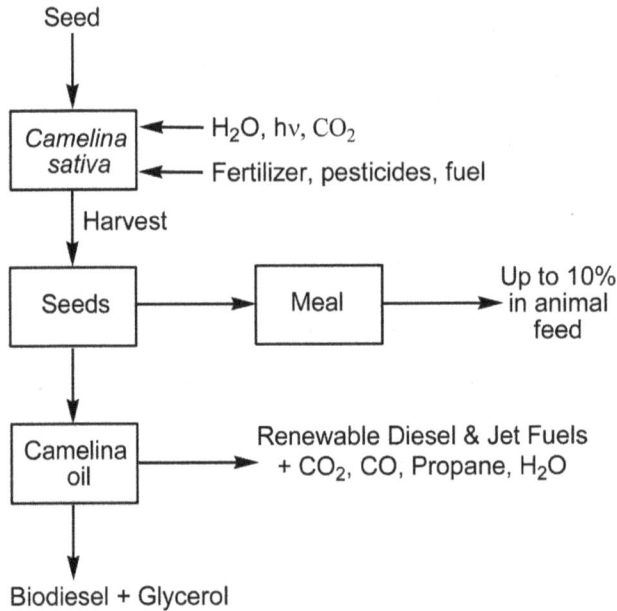

Life cycle diagram for production of alternative diesel fuels from camelina

Jatropha

Jatropha curcas Linnaeus is a species of flowering plant in the spurge family, Euphorbiaceae; the tree is shown in figure and its seeds are shown in figure. It is a tree or shrub that is native to the American tropics, mostly Mexico and Central America, but it grows under a variety of agroclimatic conditions and is commonly found in most of the tropical and subtropical regions of the world. Thus, it ensures a reasonable production of seeds with very minor care. The oil content of Jatropha seed ranges from 30% to 35% by weight. The common by-products produced while processing the biodiesel are glycerol and oilseed cake.

These by-products shall reduce the cost of biodiesel depending upon the price which these products can fetch. The cost components of biodiesel are the price of seed, seed collection and oil extraction, oil trans-esterification, transport of seed and oil. The cost of biodiesel produced by trans-esterification of oil obtained from Jatropha curcas seeds will be very close to the cost of seed required to produce the quantity of biodiesel as the cost of extraction of oil and its processing in to biodiesel is recoverable to a great extent from the income of oil cake and glycerol which are by-products.

Rapeseed

Rapeseed is related to mustard and to other cabbage-family crops. Rapeseed has been cultivated since the 20th century B.C. Because the plant can grow with less sunlight and at lower temperatures than other crops, it was cultivated in Europe as early as the 13th century A.D.

Rapeseed oil has been used for cooking, lighting, and industrial uses. However, traditional rapeseed contains high quantities of erucic acid and glucosinolates, which make the seed meal unpalatable and possibly dangerous to livestock if fed in large quantities.

The Technological Process of Obtaining Biodiesel

Preparation of Rapeseed Oil

The biodiesel is produced mainly from rapeseed with doubly improved varieties of seeds. The composition of rapeseed oil must meet specific quality requirements that are included in the standard for methyl esters as fuels for diesel engines.

Vegetable oils can be obtained by mechanical or/and chemical extraction. Crushing of seeds is a traditional way of producing oils using different types of presses. However, chemical extraction is a modern way of obtaining vegetable oils using solvent extracts (e.g. hexane). These methods can be combined e.g. residue after pressing is subjected to solvent extraction. Some smaller plants are rather only press-plants without applying the step of solvent extraction.

The process of rapeseed oil preparation consists of seed cleaning, preconditioning, flaking, cooking, screw pressing, solvent extraction, desolventizing, distillation and degumming. Rapeseed cleaning involves aspiration, screen separation to remove oversized particles, and screen separation to remove undersized particles. Next, the whole seeds are preheating prior to processing (to about 30- 40°C) by indirect heating or direct hot air contact. Cell walls must be ruptured in order to extract the oil. Moreover, it allows a solvent to get into cellular structure dissolving and diluting the lipid portions which then are separated from the solid flake. Next, the preheated rapeseed oil is flaked between two smooth surface cast-iron rolls. Extraction of oil from flaked rapeseed proceed mostly by pre-press solvent extraction. Then there is further extraction of oil seeds and press cake with hexane. In the next step, the hexane solvent is removed from the extracted cake and then distilled from the rapeseed oil. The phospholipids or gums need to be removed in a degumming step. Next, free from moisture and cooled oil is transferred to the refining process or into a storage. Crude oil obtained by extraction contain many undesirable substances such as mucoid substances, coloring matters and free fatty acids. Therefore, it has to be refined to obtain a high quality oil. The refining process involves degumming, neutralization, drying, bleaching, and deodorization. Gums compose about 2% of solvent-extracted rapeseed oil. Degumming treatment uses hot water or steam and phosphoric acid, citric acid, or other acidic materials. Precipitated gums are removed by centrifugation. Free fatty acids can be neutralized with alkali solution. After that, the oil is washed with hot water to remove traces of soaps that can reduce stability of oil. Next, the oil is dried to remove traces of water. Bleaching process proceeds by adsorption of the color producing substances on an adsorbent material such as bentonite or Fuller's earth. The next step i.e. deodorization is a vacuum steam distillation process in order to remove trace constituents giving rise to undesirable odors in oils.

The high contents of free fatty acids and water in the collected oil are responsible for secondary reactions during transesterification. Therefore, a pretreatment of oil is necessary before load the reactor to produce biodiesel. Oil refining aims to remove the excess of phospholipids, salts of iron or copper and obtain low peroxide number as well as low acid number (less than 1mg KOH/g) i.e. low content of free fatty acids which react with basic catalyst during transesterification to form soap making difficult purification of glycerin phase and increasing the demand for catalyst. The presence of phospholipids in the oil increases its resistance to oxidation, but also emulsifying reaction system makes it difficult separation of glycerol and ester phase. The phosphorus content in the oil directed to transesterification shouldn't exceed 10 ppm. Rapeseed oil led to methanolysis should be deprived of moisture (water content<0.5%), because its presence causes the hydrolysis of triacylglycerol, resulting in the formation of free fatty acids. Some factors decide about quality of rape seeds i.e. the degree of seeds maturity, the presence of damaged seeds, seed moisture content and storage times and conditions. Higher free fatty acids content can be stated in the seeds of inferior quality.

Transesterification of Rapeseed Oil

Transesterification is the best way to obtain biodiesel because it is well-known and cheap process which gives less problems for the engines than another methods. In a standard process of production biodiesel from rapeseed oil there are following process steps i.e. esterification of rapeseed oil, separation of esterification products, methanol distillation and purification of the ester. The main stage of the process is based on the transesterification reaction of rapeseed oil with an alcohol (methanol, ethanol) which results in formation of esters of alcohols and glycerol. The reaction is reversible due to formation of water, which is responsible for shifting the equilibrium towards the reagents. In order to move the chemical equilibrium towards the ester, an excess of alcohol is used. The transesterification involves three consecutive and reversible reactions. Each molecule of alcohol makes with the residue of fatty acid a distinct monoester molecule. If the triglyceride contains three different fatty acids in its molecule, a mixture of rapeseed oil esters is obtained.

There are used various chemical catalysts in the transesterification reaction i.e. acids, alkalis and enzymes. Some of them are the most effective i.e. alkaline catalysts and their methoxides. Only anhydrous reactants should be used, because the water decomposes catalyst and leads the reaction to the side of the reactants. The amount of the catalyst depends on its type, quality of substrates, reaction time and temperature. Its value varies from 0.2% to 2% by mass relative to the weight of oil. Unrefined oil requires more catalyst. Free fatty acids content in oil shouldn't exceed 0.5-1%. The higher the water content in methanol, the greater the consumption of catalyst and higher contents of free fatty acids as well as soaps. Forming of soaps reduces the activity of the catalyst and increases the viscosity, so the separation of the glycerol is more difficult.

In the majority of the biodiesel production processes, methanol is used, mainly due to the cost and its physical and chemical advantages (polar, shortest chain alcohol and the reactivity is easier). For the basic-catalysed process, the optimal ratio is 6:1 (alcohol: oil). Applying alkaline catalysts in the reaction allow to carry out the transesterification process at room temperature. The reaction time is very diverse and can vary from several minutes to several hours (typically about 30-60 minutes). Mixing is important because of the lack of mutual solubility of the substrates. Appropriately intensive mixing of reactants increases the contact area. Using the refined oils with a higher purity as well as pure and anhydrous methanol, allows to obtain biodiesel with a high content of methyl ester (96.5- 99%).

In the transesterification reaction, the oil is converted to FAME (Fatty Acid Methyl Ester). The flow, which goes out from the reactor, should be purified. First, the methanol is recovered (94%) and then residue is carried to separation of glycerin. Next, the obtained biodiesel is purified to commit the standards of quality.

The primary by-products generated during the production of biodiesel include: oilseed meal, straw and glycerol. Oilseed meal is produced in the process of pressing oil from rapeseeds. It accounts for 70% by weight, and oil represents the remaining 30%. Oilseed meal can be used as a component of feed or as a component of the binder for the production of smokeless fuel, additive to fuel briquettes and as a basic component of mat to cover mine dumps.

Straw is formed at the beginning of the production cycle. Part of straw in the form of chaff can be plowed already on the field and is a valuable organic fertilizer. Properly prepared straw can also be used as a component of the roughage. Straw is also a valuable energy source, because its calorific value is about 30-40% of the calorific value of coal and the ash formed of its combustion can be used as a mineral fertilizer.

Glycerol is a byproduct formed in the esterification of rapeseed oil. In this process also are created soaps and free fatty acids. Glycerol phase constituting concentrated glycerol (about 80% solution) may be transferred to facilities specializing in treatment of glycerol (e.g. cosmetic, pharmaceutical). The cost of obtaining the crude glycerol during esterification of vegetable oils is lower than the cost of traditional methods that had been previously used in splitting of fats. Glycerol from vegetable oils is used both for cosmetic purposes and pharmacology. More often it is an ecological material for the manufacture of certain foods and it is used for making explosives. Using own purification of glycerol, receives an additional by-product in the form of valuable mineral fertilizer. In the case of small factories, where it is uneconomic to build installation of treatment and it is not organized buying glycerol, it can be disposed of by diluting with water in a ratio of 1:100 and added to the slurry.

Soaps and free fatty acids also may be forwarded to the appropriate processing plants, or used in their own farm. Mud can be a valuable addition to the feed for producing Californian earthworms Biohumus – mineral fertilizer.

Vegetable oil is low-density-energy biomass. It limits the scale of biodiesel production. However, there are advantages of biodiesel plant refinery such as low investment and rapid construction.

The main application of fatty acid methyl esters is biodiesel, generally as mixture with petro-diesel for engine use. After refinement, fatty acid methyl esters can be also used as biodegradable solvents. Additionally, there are a source of many commodity chemicals such as higher alcohols. Glycerol is an important chemical product mainly used in areas such as food, beverage, pharmaceutical, cosmetic, tobacco and paper making.

Animal Fat

About one-third of the fats and oils produced in the United States are animal fats. This includes beef tallow, pork lard, and chicken fat. Animal fats are attractive feedstocks for biodiesel because their cost is substantially lower than the cost of vegetable oil. This is partly because the market for animal fat is much more limited than the market for vegetable oil, since much of the animal fat produced in the U.S. is not considered edible by humans.

Animal fat is currently added to pet food and animal feed, and used for industrial purposes such as soap making. Much of the domestic animal fat supply is exported.

Animal fat feedstocks can be made into high-quality biodiesel that meets the ASTM specifications for biodiesel. However, there are some drawbacks and challenges to using animal fat feedstocks.

Processing of Animal Fats

Waste fat from animal carcasses are removed and then made into an oil using a rendering process. Rendering consists of grinding the animal by-products to a fine consistency and cooking them until the liquid fat separates and pathogens are destroyed. The solids are usually passed through a screw press to complete the removal of the fat from the solid residue. The cooking process also removes water, which makes the fat and solid material stable against rancidity. The end products are fat, and a high-protein feed additive known as "meat and bone meal."

Fatty Acid Content of Animal Fats

Animal fats are highly saturated, which means that the fat solidifies at a relatively high temperature. Therefore, biodiesel made from animal fat has a high cloud point. For example, biodiesel made from beef tallow and pork lard has a cloud point in the range of 55°F to 60°F. B100 (pure biodiesel) made from animal fat should only be used in a very warm climate. However, animal fat biodiesel can be blended with petro-diesel. At lower blends such as B5 (a blend of 5% biodiesel with 95% petro-diesel), the high cloud point of the animal fat biodiesel does not have much effect on the cloud point of the blend.

The composition, or fatty acid profile, of various animal fats is shown in table below. These data show how saturated the fat will be, which in turn determines how readily it will solidify as the temperature decreases.

Table: Fatty Acid Percentages in Animal Fats

Fatty acid	Beef tallow	Pork lard	Chicken fat
Myristic 14:0	1.4 - 6.3	0.5 – 2.5	1
Palmitic 16:0	20 - 37	20 – 32	25
Palmitoleic 16:1	0.7 – 8.8	1.7 – 5	8
Stearic 18:0	6 – 40	5 – 24	6
Oleic 18:1	26 – 50	35 – 62	41
Linoleic 18:2	0.5 - 5	3 – 16	18

Beef tallow and pork lard are typically about 40% saturated (sum of myristic, palmitic and stearic acids). Chicken fat is lower at about 30-33%. For comparison, soybean oil is about 14% saturated and canola oil is only 6%. Thus, tallow and lard are usually solid at room temperature and chicken fat, while usually still liquid, is very viscous and nearly solid.

When the animal fat is made into biodiesel, the concern about solidification at lower temperatures continues. The methyl esters from the saturated fatty acids, mainly methyl stearate and methyl palmitate, have high melting points.

Oxidative Stability of Animal Fat Biodiesel

In theory, the saturated fatty acids in animal fats should contribute to better oxidative stability for biodiesel. Animal fats contain very little of the polyunsaturated fatty acids, such as linoleic acid and linolenic acid, that make vegetable oils such as soybean oil and linseed oil so prone to rancidity.

However, in practice animal fat is not always more stable than vegetable oil, because vegetable oils often contain natural anti-oxidants. For example, a test of the peroxide content of lard and vegetable oil showed that the lard experienced oxidation faster than the vegetable oil.

Animal fats contain very little of the natural anti-oxidants such as vitamin E that protect vegetable oils. In some cases, used cooking oils may also contain artificial antioxidants that are added to oil to extend their life, and these artificial anti-oxidants can in turn extend the life of biodiesel made from used cooking oil.

High Cetane Number of Animal Fat Biodiesel

Animal fat feedstocks result in biodiesel with a high cetane number, which is an important quality parameter for diesel fuels. The saturated fatty acids are the source of this high cetane number and values over 60 are common. Soybean oil based biodiesel usually has a cetane number of about 48-52 and petroleum-based diesel fuel is usually between 40 and 44. When animal fat biodiesel is blended with petro-diesel, this high cetane number can help the engine start more quickly and run more quietly.

Lower Nitrous Oxide Emissions from Animal Fat Biodiesel

One of the important attributes of biodiesel is that it lowers the levels of harmful pollutants in the exhaust of diesel engines. One exception to this is oxides of nitrogen (NOx), which are implicated in ozone and smog formation. Biodiesel tends to emit slightly more nitrous oxides than petro-diesel. Many reasons have been cited for this increase in NOx and there is still considerable debate.

However, it has been shown that biodiesel from animal fats tends to produce a smaller increase in NOx and in some cases no increase. The primary reason for this is probably that animal fat biodiesel has a high cetane number (>60) compared with vegetable oil biodiesel. Higher cetane number is known to lower NOx by lowering temperatures during the critical early part of the combustion process.

Contaminants in Animal Fat Feedstocks

Animal fats can be challenging feedstocks for biodiesel because they frequently contain contaminants that should be removed before the fuel is used in an engine.

Phospholipids, or gums, will cause insoluble precipitates when they come into contact with water. Since these precipitates will plug fuel filters, they must be removed from the fuel. Most of this material will separate with the glycerin during processing, or be removed during the purification step (water washing or ion exchange), so it is rare for these compounds to be found in the fuel. In fact, since phospholipids can de-activate exhaust aftertreatment devices on diesel vehicles, they are usually removed from the feedstock before it is converted to fuel. Most commonly, the gums are removed by adding water and citric or phosphoric acid to the feedstock and then separating the precipitates with a centrifuge.

Polymers that are formed naturally at the high temperatures of the rendering process can contribute to a higher viscosity in biodiesel made from animal fat. This higher viscosity can sometimes prevent animal fat biodiesel from passing the viscosity ASTM specification. If this is the case, the animal fat biodiesel can be blended with a lower-viscosity biodiesel.

An occasional problem with rendered fats is polyethylene in the fat which originates from plastic bags, ear tags, or other plastic that is mixed with the animal byproducts. Finely divided polyethylene has been found to cause cloudiness in the fuel and could plug the fuel filter. Producers who are used to working with vegetable oil may be confused by this, since with vegetable oil biodiesel, cloudiness is caused mainly by water in the fuel. If animal fat biodiesel shows cloudiness even after water is removed, this may be caused by stray polyethylene, which can be removed by passing the biodiesel through a fine filter.

High Sulfur Content of Animal Fat Biodiesel

Sulfur can sometimes be a problem for animal fat-based biodiesel. Biodiesel sold for on-highway use is only allowed to contain up to 15 ppm of sulfur. Some samples of beef tallow have been found to contain over 100 ppm of sulfur and chicken fat frequently contains a similar amount. The sulfur apparently originates from sulfur-containing amino acids associated with proteins that carry over from the rendering process.

Measurements of the sulfur levels in biodiesel produced from animal fats have shown that the sulfur level usually decreases by about half when the conversion takes place. However, the remaining sulfur can be difficult to remove. Vacuum distillation is about the only reliable technique for removing this sulfur.

Alternatively, biodiesel high in sulfur can be sold for non-highway use, such as fuel for a boiler or heater.

Paper Waste

Recycled paper, paper sludge from production and even sawdust from the early processing stages of paper have all been examined as possible sources of biofuel. While producers would have to compete with paper companies -- who use the waste in their own recycling programs --researchers believe that some waste could be diverted for fuel production. However, many of these materials might be more trouble than they are worth. that paper is difficult to convert to liquid fuel because of its waxy coating and how the paper is made. "Have you ever tried to light up a newspaper?"."It will burn but then goes out right away because of the form of the paper. It's a complex system; it's a lot harder [to burn] than wood or corn stalks." Paper waste rounds out the bottom of the countdown because it's a source that is already being widely used for products other than biofuel. Combined with the fact that it is difficult to process and somewhat expensive makes this biofuel a limited opportunity.

Materials and Methods

Collection of Substrate

Newspaper, whichwas used as a substrate for the production of bioethanol,was collected from the households.The substrate was collected in a dust free and fungus-free state and was dried in sunlight and was made into small pieces and stored in sealed plastic bags.

Chemical Analysis of the Substrate

Composition of the substrate and its properties were analyzed before pre-treatment. The cellulose content and total carbohydrate in the substrate was estimated by anthrone method, Moisture content and ash content of the substrate were also estimated using standard methods.

Optimization of Pretreatment Process

The pre-treatment optimization for the substrate was carried out by using different combination dilute sulphuric acid ranging from 0 to 6% and heating period of 30, 45 and 60 minutes at 1210 C and 15lb pressure. 1gm of substrate was added with 10 ml of dilute sulphuric acid (1:10). Cellulose released during this optimization process was analysed by anthrone method. After the release of maximum amount of cellulose during pre-treatment process, the solution was taken for hydrolysis.

Hydrolysis of the Pretreated Substrate

Maximum cellulose released during the pretreatment was hydrolysed by the isolated cellulose degrading bacteria. The pretreated substrate was washed with distilled water several times to neutralise the acid concentration. The substrate was oven dried till constant weight and the pH was adjusted to 7.0. Comparison study between isolated cellulose degrading bacteria and the pure culture, CH was performed. A 24hr grown inoculum of isolated cellulose degrading bacteria and pure culture, CII were added to the pretreated substrate. Reducing sugars release during substrate hydrolysis were analysed by Dinitrosalicylic Acid (DNS) method every 24hr from zero hour, for both the organisms. Maximum sugars released during this period were further taken for fermentation to produce bioethanol.

Fermentation of Hydrolysed Broth

Fermentation was carried out using commercially available yeast, Saccharomyces cerevisiae. The pH of hydrolysed broth was adjusted to 4.6 and an inoculum of active yeast (in log phase) was added to the hydrolysed broth. The fermentation was carried out at 360 C until maximum sugars are converted into bioethanol. The reducing sugar utilization during fermentation was analysed by DNS method and the bioethanol production was analysed by using specific gravity method.

Calculation for specific gravity:

$$\frac{W2 - W1=}{W3 - W1} \; Specific \; Gravity$$

Where,

W1 = empty weight of specific gravity bottle

W2 = Weight of sample + specific gravity bottle

W3 = Weight of distilled water + specific gravity bottle.

Also the ethanol yield was estimated using High Performance Liquid Chromatography (HPLC). The instrument used was "SHIMADZU-C-10AVP".Ethanol was analyzed by using HPLC with cation exchanger SugarPak column (C18). Acetone-Nitrate and water (80:20) was used as a mobile phase at a flow rate of 1mL/min. The injection volume was 5µl and the column temperature was maintained at 90°C. All the samples were filtered through a 0.45µm filter before subjected to HPLC analysis. The evaluate was detected by a refractive index detector at 50°C.

Results and Discussions

Chemical Analysis of the Substrate

The chemical analysis of the substrate showed that the substrate consisted of 45% cellulose. Ash content and moisture content were found to be 5.17% and 6% respectively.

Optimization of Pretreatment of the Substrate

In pre-treatment with dilute sulphuric acid, the structure of the cellulosic biomass will be altered to make cellulose more accessible to the enzymes that convert the carbohydrate polymers into fermentable sugars rapidly and with greater yield.

References

- Corn-for-biofuel-production-27536: extension.org, Retrieved 15 April 2018

- Sugarcane-biomass: bioenergyconsult.com, Retrieved 10 May 2018

- Rapeseed-and-canola-for-biodiesel-production-26629: extension.org, Retrieved 22 July 2018

- Animal-fats-for-biodiesel-production-30256: extension.org, Retrieved 19 March 2018

- Top-10-sources-for-biofuel-1769457447: seeker.com, Retrieved 31 March 2018

Chapter 5

Chemistry and Production of Biofuels

In order to completely understand biofuel chemistry and production, it is vital to understand the processes related to it. The following chapter elucidates the various methods of production of biodiesel, bioethanol, biogas and aviation biofuel as well as includes a detailed discussion of ultra- and high-shear in-line and batch reactors, supercritical process and ultrasonic reactor method.

Biofuel Chemistry

Biofuel Chemistry examines the combustion chemistry of compounds that constitute typical biofuels, including alcohols, ethers and esters. Biofuel is associated with only a few select chemical compounds, especially ethanol (used exclusively as a gasoline replacement in spark-ignition engines) and very large methyl esters in biodiesel (used as a diesel fuel replacement in diesel engines). The biofuels are oxygenated fuels, which distinguishes them from hydrocarbons in conventional petroleum-based fuels.

The Chemistry of Biodiesel

Differences between Biodiesel, Diesel and Vegetable Oil

A typical molecule of biodiesel looks like the structure below. Mostly it is a long chain of carbon atoms, with hydrogen atoms attached, and at one end is what we call an ester functional group (shown in blue).

Diesel engines can burn biodiesel fuel with no modifications (except for replacing some rubber tubing that may soften with biodiesel). This is possible because biodiesel is chemically very similar to regular diesel, shown below. Notice that regular diesel also has the long chain of carbon and hydrogen atoms, but doesn't have the ester group shown in blue above.

Actually, the first diesel engines didn't run on "diesel" fuel, but on vegetable oil, a sample molecule of which is shown below. Notice that it also has the long rows of carbon and hydrogen atoms, but is about three times larger than normal diesel molecules. It also has ester functional groups (in blue), like biodiesel.

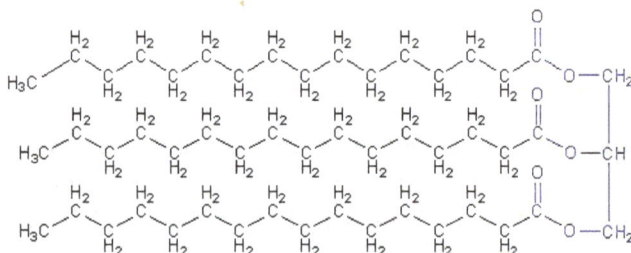

That larger size of vegetable oil means that in cold weather it gels, making it hard to use in an engine. Converting it into biodiesel makes it a smaller molecule, closer to the size of regular diesel, so that it has to get colder than vegetable oil before it starts to gel.

Chemical Conversion of Vegetable Oil to Biodiesel

Vegetable oil, like biodiesel, belongs to a category of compounds called esters. Therefore, converting vegetable oil into biodiesel is called a transesterification reaction. Doing this reaction requires using methanol (shown in green), which causes the red bonds in the structure below to break. This breaks off the blue section, like a backbone on the molecule, which becomes glycerol. The red bonds that did go to the glycerol backbone are placed with bonds to methoxy groups, shown in green in the final structure, that came from the methanol:

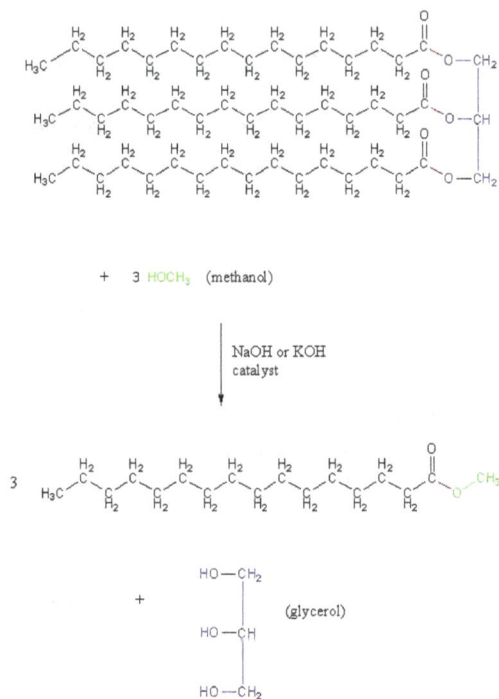

Other Steps in Making Biodiesel

Making good biodiesel requires several other steps besides the transesterification reaction. The first is to remove any traces of water in the vegetable oil. If this is not done, the water will later react with the vegetable oil in the reaction and make soap, shown below.

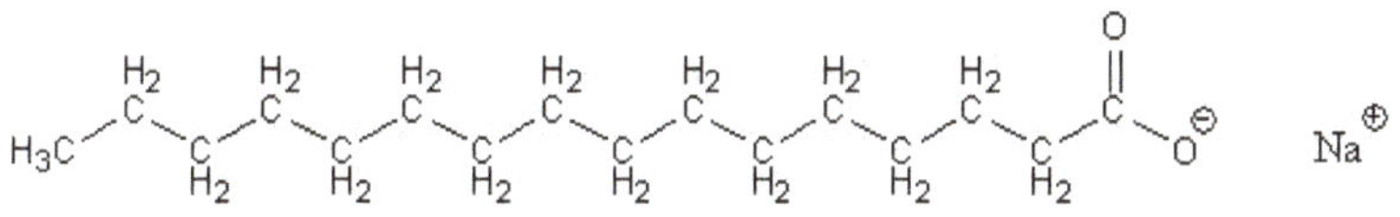

If soap gets made, then later it complicates the steps after the transesterication reaction that are needed to separate the biodiesel from leftover methanol, the NaOH or KOH catalyst, and the glycerol byproduct.

Ethanol, Butanol and Gasoline

The biggest difference between these biofuels (ethanol and butanol) and biodiesel, is that organisms can produce ethanol and butanol on their own. This means there is no need to harvest an organism to obtain a raw product that is converted to biofuel. The advantage of such an approach is that fewer nutrients and less water are required.

The molecular structure of ethanol, butanol, and gasoline are diagramed below. Note that the only real difference, structurally, between ethanol and butanol is the number of carbon atoms they contain. Note that octane is used to represent gasoline even though gasoline is actually a mixture of hydrocarbon molecules ranging in size from four to twelve carbon atoms long.

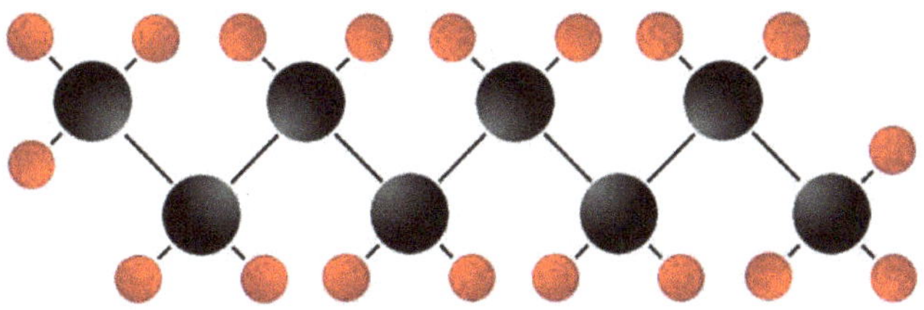

Octane, one of several molecules in gasoline.

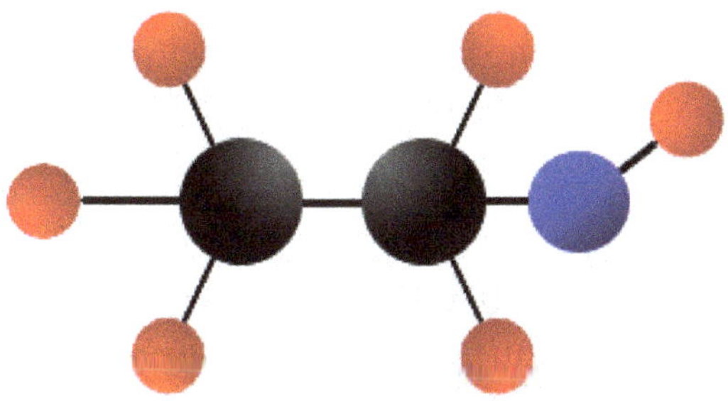

Ethanol (blue represents oxygen)

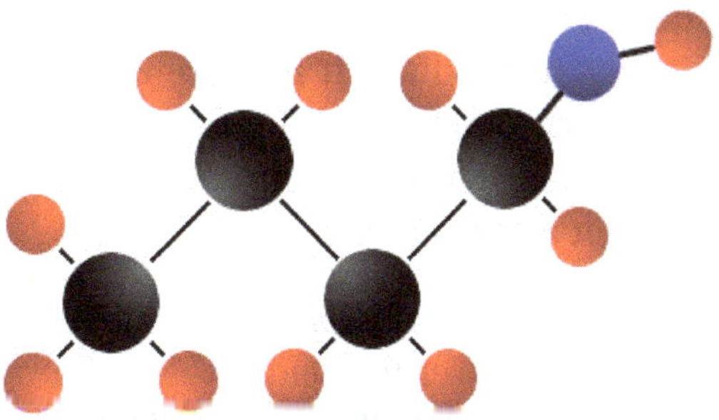

Butanol (blue represents oxygen)

Ethanol and butanol are both considered biofuels because they can be produced from living or recently living organisms. Like the differences between petrodiesel and biodiesel, the big difference between ethanol/butanol and gasoline is the presence of oxygen.

In the case of alcohols (ethanol and butanol are alcohols) there is no ester group, but there is an alcohol group (just an oxygen with a hydrogen attached). This difference means that alcohols behave differently from the esters that make up biodiesel. The chart below compares gasoline to ethanol, butanol, and biodiesel.

Property	Gasoline	Ethanol	Butanol	Biodiesel
Octane (average)	85-96	99.5	97	(cetane) 50-65
Energy Density (MJ/kg)	33	20	30	38

The thing to note is that ethanol has a very low energy density compared to gasoline, but butanol is pretty comparable. It requires about 1.5 times as much ethanol as it does gasoline to generate the same power. In a car, that means you need 1.5 gallons of ethanol to travel the same distance as you would on one gallon of gasoline.

Ethanol versus Butanol

Clearly butanol is more similar to gasoline than is ethanol. The problem is that producing butanol is more difficult and has not been reliably demonstrated on a large scale. Many organisms, like yeast and even some types algae, produce ethanol as part of their normal chemistry. That is to say, a number of organisms produce ethanol just like plants produce oxygen or humans produce carbon dioxide. Many organisms only need carbon dioxide and an energy source to produce ethanol. In the case of algae, only water, sunlight, and CO_2 are needed. Thus, it is easy to produce ethanol.

The problem with ethanol, beyond energy density, is that it is highly corrosive to engine component, particularly rubbers, and it takes a great deal of energy to produce. Changing feedstock from corn to something like Jatropha or simply using algae can help to offset this energy input, but so far it seems that more energy is needed to produce ethanol than is derived from it. Besides, algae are better at producing biodiesel.

Butanol is less corrosive than ethanol and because it takes about the same amount of energy to produce, it is more efficient in the long run. The problem with butanol production is that it is difficult to do on large scale. Recently, several companies have devised plans to produce butanol in large quantities and are moving forward with plans to make it a commercially viable fuel.

Like biodiesel, both ethanol and butanol are biodegradable. Unlike biodiesel, they do not freeze at lower temperatures and are more stable during long term storage. Thus, these biofuels may offer better alternatives to gasoline, in the long run, than biodiesel offers compared to petrodiesel.

Burning of Biofuels

Burning Alcohol

Ethanol, methanol, butanol, and other alcohols are all flammable. Interestingly, both alcohols and gasoline don't burn in their liquid forms. That is to say, the liquid forms of these molecules are relatively difficult to burn, why is that?

The answer has to do with oxygen (molecular oxygen). Combustion requires oxygen and only gasoline or alcohol exposed to oxygen can burn. In the liquid form, both substances are packed tight enough to prevent too much oxygen from entering the liquid. This is why gasoline burns better when sprayed into an engine by a fuel injector. It is also why the surface of ethanol burns and not the entire liquid contents. So, both fuels need to be aerosolized in order to burn efficiently.

When ethanol is burned in the presence of oxygen, we get an equation that looks something like this.

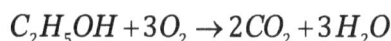

$$C_2H_5OH + 3O_2 \rightarrow 2CO_2 + 3H_2O$$

Combustion of Ethanol

When butanol is burned, we see this.

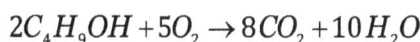

$$2C_4H_9OH + 5O_2 \rightarrow 8CO_2 + 10H_2O$$

Combustion of Butanol

These equations are, of course, the ideal relationships. Because alcohol contains very little sulfur, there is very little sulfur dioxide produced when it is burned and thus little sulfuric acid. However, alcohol is not a "clean-burning fuel." Alcohol, and any fuel derived from recently living plant and animal matter, contains larger amounts of nitrogen. This means that biofuels produce more nitric oxide and other nitrogen compounds when burned. In some cases, this not only offsets the savings from not producing sulfur compounds, but actually worsens the long term trend for acid rain.

Alcohols are not immune to inefficient combustion issues either. The inefficient combustion of alcohol produces carbon monoxide, formaldehyde, ammonia, benzene, and other toxic chemicals. In some studies, the combustion of ethanol actually produced more formaldehyde, a toxic and carcinogenic compound, than burning gasoline. Ultimately, burning ethanol does produce less carbon monoxide because the fuel itself supplies some oxygen in addition to what is found in the atmosphere. The benefits in terms of carbon monoxide are less with butanol because the ratio of oxygen to carbon is lower in butanol. In ethanol, the ratio is one oxygen molecule to two carbon molecules or 1:2. In butanol, the ratio is 1:4.

We can do a bit of analysis to determine how much byproduct ethanol produces in comparison to gasoline. This comparison will only be for the actual combustion step and does include production, distribution, etc.

When a kilogram of gasoline is burned, we get approximately 33 megajoules of energy and somewhere on the order of 3.09 kilograms of carbon dioxide. When a kilogram of ethanol is burned, we get 20 megajoules of energy and about 1.91 kilograms of carbon dioxide. However, remember that we get 20/33 times the amount of energy from ethanol or only about 0.61 times the amount of energy. This means we need to burn 1.64 kilograms of ethanol to get the equivalent amount of energy, which brings us up to 3.14 kilograms of carbon dioxide. So, ethanol produces MORE carbon dioxide than gasoline.

For butanol, about 2.37 kilograms of carbon dioxide are produce per kilogram burned. Because it returns somewhere on the order of 29 MJ/kilogram, we only need about 1.14 kilograms to get the same energy content as gasoline. Thus, for the same energy content, butanol produces only about 2.7 kg of carbon dioxide, putting it ahead of gasoline and well ahead of ethanol.

Burning Biodiesel

There are two types of biodiesel that are generally used when calculating combustion, the C19 and the C20 chain lengths. We will focus just on C19 (nineteen carbons in the chain) to keep things simple. The equation for ideal combustion of biodiesel looks like this.

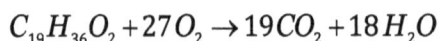

$$C_{19}H_{36}O_2 + 27O_2 \rightarrow 19CO_2 + 18H_2O$$

Ideal Combustion of Biodiesel

If we run the math on this equation, we find that biodiesel produces about 2.52 kilograms of carbon dioxide for every kilogram of fuel burned (2.59 if we use C20). This compares very favorably to petrodiesel, which produces 3.17 kilograms of carbon dioxide per kilogram of fuel. If you figure in the reductions in sulfur emissions, then biodiesel seems to be faring quite well (at least better than ethanol). Of course, we cannot forget to include the energy conversion, which is about 38 MJ/kg for biodiesel and about 43 MJ/kg for petrodiesel. Thus, we need about 1.13 times as much biodiesel to get the same amount of energy and will therefore produce around 2.86 kg of carbon dioxide.

The biggest problem with biodiesel is the production of nitrogen compounds, namely nitric oxide, which are poisonous in and of themselves and contribute to the production of acid rain. In a strange twist of irony, it has been shown that varying the temperature in a diesel engine can reduce the production of nitrogen compounds only to increase the production of soot. Thus, there is a tradeoff that seems to greatly limit the utility of biofuel in reducing overall emissions.

Biofuel Production

Production of Biodiesel

The raw materials for biodiesel production are vegetable oils, animal fats and short chain alcohols. The oils most used for worldwide biodiesel production are rapeseed (mainly in the European Union countries), soybean (Argentina and the United States of America), palm (Asian and Central American countries) and sunflower, although other oils are also used, including peanut, linseed, safflower, used vegetable oils, and also animal fats. Methanol is the most frequently used alcohol although ethanol can also be used.

Since cost is the main concern in biodiesel production and trading (mainly due to oil prices), the use of non-edible vegetable oils has been studied for several years with good results.

Besides its lower cost, another undeniable advantage of non-edible oils for biodiesel production lies in the fact that no foodstuffs are spent to produce fuel. These and other reasons have led to medium- and large-scale biodiesel production trials in several countries, using non-edible oils such as castor oil, tung, cotton, jojoba and jatropha. Animal fats are also an interesting option, especially in countries with plenty of livestock resources, although it is necessary to carry out preliminary treatment since they are solid; furthermore, highly acidic grease from cattle, pork, poultry, and fish can be used.

Microalgae appear to be a very important alternative for future biodiesel production due to their very high oil yield; however, it must be taken into account that only some species are useful for biofuel production.

Although the properties of oils and fats used as raw materials may differ, the properties of biodiesel must be the same, complying with the requirements set by international standards.

Characteristics of Oils and Fats used in Biodiesel Production

Oils and fats, known as lipids, are hydrophobic substances insoluble in water and are of animal or vegetal origin. They differ in their physical states at room temperature. From a chemical viewpoint, lipids are fatty glycerol esters known as triglycerides. The general chemical formula is shown in figure below.

In the figure below, R_1, R_2 y R_3 represent hydrocarbon chains of fatty acids, which in most cases vary in length from 12 to 18 carbon atoms. The three hydrocarbon chains may be of equal or different lengths, depending on the type of oil; they may also differ on the number of double-covalent bonds in each chain. Fatty acids may be saturated fatty acids (SFA) or non-saturated fatty acids (NSFA). In the former, there are only single covalent bonds in the molecules. The

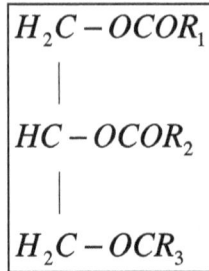

$$
\begin{array}{l}
H_2C - OCOR_1 \\
\quad | \\
HC - OCOR_2 \\
\quad | \\
H_2C - OCR_3
\end{array}
$$

General chemical formula of triglycerides

Table: Chemical formulas of the main fatty acids in vegetable oils

Fattyacid	Chemical formula
Lauric	(12:0) CH3 (CH2)10 COOH
Palmitic	(16:0) CH3 (CH2)14 COOH
Estearic	(18:0) CH3 (CH2)16 COOH
Oleic	(18:1) CH3 (CH2)7 CH = CH (CH2)7 COOH
Linoleic	(18:2) CH3 (CH2)4 CH = CH CH2 CH = CH (CH2)7 COOH
Linolenic	(18:3) CH3 CH2 (CH = CH CH2)3 (CH2)6 COOH
Erucic	(22:1) CH3 (CH2)7 CH = CH (CH2)11 COOH
Ricinoleic	(18:1) CH3 (CH2)5 CHOH CH2 CH = CH (CH2)7 COOH

Table: Approximate content (in weight) of saturated and non-saturated fatty acids in some vegetable oils and animal fats

Oil/fat	SFA (& % w/w)	NSFA (& % w/w)
Coconut	90	10
Corn	13	87
Cottonseed	26	74
Olive	14	86
Palm	49	51
Peanut	17	83
Rapeseed	6	94
Soybean	14	86
Sunflower	11	89
Safflower	9	91
Castor	2	98

Yellow grease	33	67
Lard	41	59
Beef tallow	48	52

names of the most important fatty acids in oils are listed in first table along with their chemical formulas. The notation x:y indicates the number of carbon atoms in the oil molecule (x) and the number of unsaturations, i.e. double-covalent bonds (y). For instance, y = 0 for all the SFAs. Table above indicates the approximate content (in weight) of saturated and non-saturated fatty acids in some vegetable oils and animal fats.

The most frequent fatty acids in oils are lauric, palmitic, estearic, linoleic and linolenic, although others may also be present. It is important to note that vegetable oils differ in their content of fatty acids. For instance, ricinoleic acid is the main component of castor oil, whereas in olive oil it is oleic acid, in soybean oil it is linoleic acid, and in linseed oil it is linolenic acid.

The compositions indicated in table above do not discriminate between the different saturated or unsaturated fatty acids. For instance, coconut oil has about 90% of SFAs in its composition (more than half being lauric acid), and palm oil has

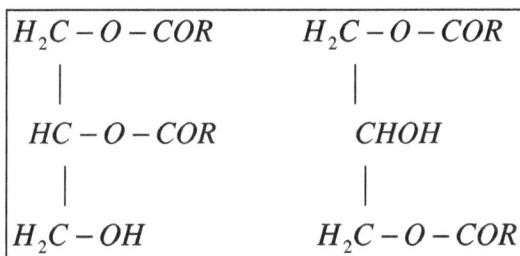

$$
\begin{array}{cc}
H_2C - O - COR & H_2C - O - COR \\
| & | \\
HC - O - COR & CHOH \\
| & | \\
H_2C - OH & H_2C - O - COR
\end{array}
$$

Chemical formula of diglycerides

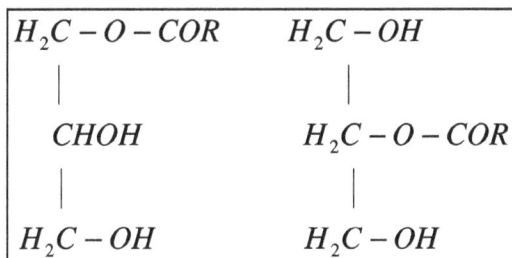

$$
\begin{array}{cc}
H_2C - O - COR & H_2C - OH \\
| & | \\
CHOH & H_2C - O - COR \\
| & | \\
H_2C - OH & H_2C - OH
\end{array}
$$

Chemical formula of monoglycerides

about 49% SFAs (more than 80% palmitic acid). Similarly, 60% of NSFAs content in soybean oil is linoleic acid, while in peanut more than 50% is oleic.

The US Department of Energy indicates that a perfect biodiesel should only comprise mono-unsaturated fatty acids.

Vegetable oils may also contain small percentages of monoglycerides and diglycerides. Their chemical formulae are shown in above two figures. In addition, there will also be small amounts of free fatty acids (in most vegetable oils, less than 1%, except for palm oil, where they can reach up to 15%).

The composition of vegetable oils influences their properties. For instance, the pour point and cloud point temperatures, cetane number and the iodine index depend on the number of unsaturations and the length of the fatty acid chains. A higher content of double-covalent bonds gives a lower solidification point and a higher iodine index.

Characteristics of Alcohols used in Biodiesel Production

Alcohols that can be used in biodiesel production are those with short chains, including methanol, ethanol, butanol, and amylic alcohol. The most widely used alcohols are methanol (CH_3OH) and ethanol (C_2H_5OH) because of their low cost and properties. Methanol is often preferred to ethanol in spite of its high toxicity because its use in biodiesel production requires simpler technology; excess alcohol may be recovered at a low cost and higher reaction speeds are reached.

It must be remembered that in order for biodiesel to be a fully renewable fuel, it should be obtained from vegetable oils and animal fats, together with an alcohol that is produced from biomass, such as bioethanol, instead of being a petrochemical product. Several countries are carrying out research towards this objective, such as Spain and Brazil.

Biodiesel Production Process

Biodiesel is produced from vegetable oils or animal fats and an alcohol, through a transesterification reaction. This chemical reaction converts an ester (vegetable oil or animal fat) into a mixture of esters of the fatty acids that makes up the oil (or fat). Biodiesel is obtained from the purification of the mixture of fatty acid methyl esters (FAME). A catalyst is used to accelerate the reaction.

According to the catalyst used, transesterification can be basic, acidic or enzymatic, the former being the most frequently used.

A generic transesterification reaction is presented in Eq.; RCOOR' indicates an ester, R"OH an alcohol, R'OH another alcohol (glycerol), RCOOR" an ester mixture and cat a catalyst:

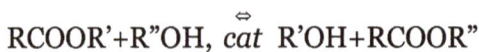

$$RCOOR' + R"OH, \overset{\Leftrightarrow}{cat} \ R'OH + RCOOR"$$

When methanol is the alcohol used in the transesterification process, the product of the reaction is a mixture of methyl esters; similarly, if ethanol were used, the reaction product would be a mixture of ethyl esters. In both cases, glycerin will be the co-product of the reaction. This is shown schematically in figures below.

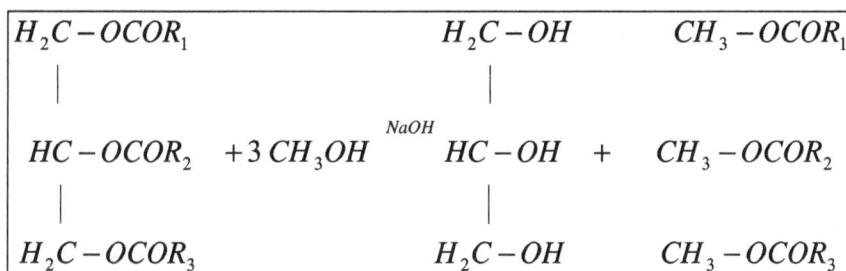

$$
\begin{array}{lll}
H_2C-OCOR_1 & H_2C-OH & CH_3-OCOR_1 \\
| & | & \\
HC-OCOR_2 \quad +3\,CH_3OH \xrightarrow{NaOH} HC-OH \ + \ CH_3-OCOR_2 \\
| & | & \\
H_2C-OCOR_3 & H_2C-OH & CH_3-OCOR_3
\end{array}
$$

Basic transesterifi-cation reaction with methanol

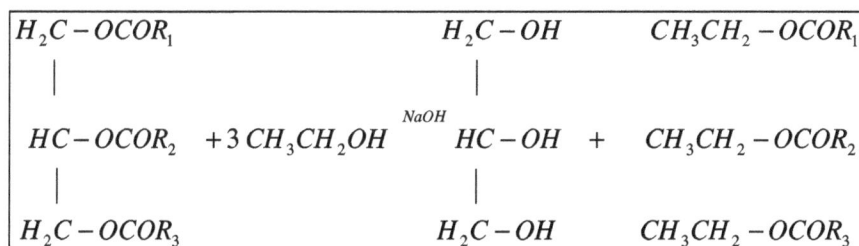

$$
\begin{array}{lll}
H_2C-OCOR_1 & H_2C-OH & CH_3CH_2-OCOR_1 \\
| & | & \\
HC-OCOR_2 \quad +3\,CH_3CH_2OH \xrightarrow{NaOH} HC-OH \ + \ CH_3CH_2-OCOR_2 \\
| & | & \\
H_2C-OCOR_3 & H_2C-OH & CH_3CH_2-OCOR_3
\end{array}
$$

Basic transesterification reaction with ethanol

Although transesterification is the most important step in biodiesel production (since it originates the mixture of esters), additional steps are necessary to obtain a product that complies with international standards. In consequence, once the chemical reaction is completed and the two phases (mix of esters and glycerin) are separated, the mix of methyl esters must be purified to reduce the concentration of contaminants to acceptable levels. These include remnants of catalyst, water and methanol; the latter is usually mixed in excess proportion with the raw materials in order to achieve higher conversion efficiency in the transesterification reaction.

Treatment of Raw Materials

The content of free fatty acids, water and non-saponificable substances are key parameters to achieve high conversion efficiency in the transesterification reaction.

The use of basic catalysts in triglycerides with high content of free fatty acids is not advisable, since part of the latter reacts with the catalyst to form soaps. In consequence, part of the catalyst is spent, and it is no longer available for transesterification. In summary the efficiency of the reaction diminishes with the increase of the acidity of the oil; basic transesterification is viable if the content of free fatty acids (FFAs) is less than 2%. In the case of highly acidic raw materials (animal fats from cattle, poultry, pork; vegetable oils from cotton, coconut, most used oils, etc.) an acid transesterification is necessary as a preliminary stage, to reduce the level of FFAs to the above-mentioned value.

Besides having low humidity and acid content, it is important that the oil presents a low level of non-saponificable substances. If the latter were to be present in significant amounts and soluble in biodiesel, it would reduce the level of esters in the product, making it difficult to comply with the minimum ester content required by the standards.

The AOCS standards list the required properties of oils. Anyway, the properties required by the oils are finally determined by the biodiesel industry in each country. For instance, in Argentina the oils for biodiesel production usually have:

- Acidity level <0.1 mg KOH/g

- Humidity <500 ppm

- Peroxide index <10 meq/kg

- Non-saponificable substances <1%.

Alcohol-catalyst Mixing

The alcohol used for biodiesel production must be mixed with the catalyst before adding the oil. The mixture is stirred until the catalyst is completely dissolved in the alcohol. It must be noted that the alcohol must be water-free (anhydrous) for the reasons explained in the previous paragraph.

Sodium and potassium hydroxides are among the most widely used basic catalysts. For production on an industrial scale, sodium or potassium methoxides or methylates are commercially available.

Of course, due caution must be exercised, and all applicable safety regulations must be followed, when working with methanol, hydroxides and methoxides, independently of the production scale.

The alcohol-to-oil volume ratio, R, is another key variable of the transesterification process. The stoichiometric ratio requires 1 mol of oil to react with 3 mol of alcohol, to obtain 3 mol of fatty acids methyl esters (FAME) and 1 mol of glycerin. However, since the reaction is reversible, excess alcohol as a reactant will shift the equilibrium to the right side of the equation, increasing the amount of products (as it may be inferred from Le Chatelier's principle). Although a high alcohol-to-oil ratio does not alter the properties of FAME, it will make the separation of biodiesel from glycerin more difficult, since it will increase the solubility of the former in the latter. Usually, a 100% alcohol excess is used in practice, that is, 6 mol of alcohol per mole of oil. This corresponds to a 1:4 alcohol-to-oil volume ratio ($R = 0.25$). Finally, it must be noted that the necessary amount of catalyst is determined taking into account the acidity of the oil, by titration.

Chemical Reaction

The chemical reaction takes place when the oil is mixed with the alkoxide (alcohol–catalyst mix) described in the previous paragraph. This requires certain conditions of time, temperature and stirring. Since alcohols and oils do not mix at room temperature, the chemical reaction is usually carried out at a higher temperature and under continuous stirring, to increase the mass transfer between the phases.

Usually, emulsions form during the course of the reaction; these are much easier and quicker to destabilize when methanol is used, in comparison to ethanol. Due to the greater stability of emulsions formed, difficulties arise in the phase separation and purification of biodiesel when ethanol is used in the reaction.

The transesterification process may be carried out at different temperatures. For the same reaction time, the conversion is greater at higher temperatures. Since the boiling point of methanol is approximately 68C (341 K), the temperature for transesterification at atmospheric pressure is usually in the range between 50 and 60C.

It is very useful to know the chemical composition of the mixture during the reaction; then, if the reaction mechanism and kinetics are known, the process can be optimized. However, the determination of the mixture composition is not easy, since more than a hundred substances are known to be present. For instance, for biodiesel production from rapeseed oil (whose main SFAs are palmitic, oleic, linoleic and linolenic) and methanol, with potassium hydroxide as a catalyst, it could be theoretically possible to find 64 isomers of triglycerides, 32 diglycerides, 8 monoglycerides, their methyl esters, potassium salts of the fatty acids, potassium methoxide, water, etc.

The studies on this subject indicate the following general guidelines:

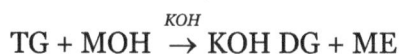

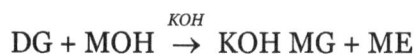

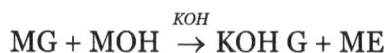

$$TG + MOH \xrightarrow{KOH} KOH\ DG + ME$$

$$DG + MOH \xrightarrow{KOH} KOH\ MG + ME$$

$$MG + MOH \xrightarrow{KOH} KOH\ G + ME$$

where MOH indicates methanol, ME are the methyl esters, TG, DG and MG are tri-, di- and monoglycerides, respectively, and G is the glycerin.

Several methods, with different levels of equipment complexity and training requirements, have been devised to analyze samples that are mixtures of fatty acids and esters from mono-, di-, and triglycerides obtained from transesterification of vegetable oils.

It must be noted that thin layer chromatography (TLC) provides essentially qualitative information about the sample composition, as distinct from the other methods, that can be used for quantitative analysis. However, the simplicity, speed and low cost of TLC make it quite attractive as a technique for process optimization and routine checks, especially in small- and medium-scale production plants, and also for training purposes.

Catalysts

The catalysts used for the transesterification of triglycerides may be classified as basic, acid or enzymatic.

Basic catalysts include sodium hydroxide (NaOH), potassium hydroxide (KOH), carbonates and their corresponding alcoxides (for instance, sodium methoxide or ethoxide). There are many references on basic catalysts in the scientific literature.

Acid catalysts include sulfuric acid, sulfonic acids and hydrochloric acid; their use has been less studied.

Heterogeneous catalysts that have been considered for biodiesel production include enzymes, titanium silicates, and compounds from alkaline earth metals, anion exchange resins and guanidines in organic polymers. Lipases are the most frequently used enzymes for biodiesel production.

Separation of the Reaction Products

The separation of reaction products takes place by decantation: the mixture of fatty acids methyl esters (FAME) separates from glycerin forming two phases, since they have different densities; the two phases begin to form immediately after the stirring of the mixture is stopped. Due to their different chemical affinities, most of the catalyst and excess alcohol will concentrate in the lower phase (glycerin), while most of the mono-, di-, and triglycerides will concentrate in the upper phase (FAME). Once the interphase is clearly and completely defined, the two phases may be physically separated. It must be noted that if decantation takes place due to the action of gravity alone, it will take several hours to complete. This constitutes a "bottleneck" in the production process, and in consequence the exit stream from the transesterification reactor is split into several containers. Centrifugation is a faster, albeit more expensive alternative.

After the separation of glycerin, the FAME mixture contains impurities such as remnants of alcohol, catalyst and mono-, di-, and triglycerides. These impurities confer undesirable characteristics to FAME, for instance, increased cloud point and pour point, lower flash point, etc. In consequence a purification process is necessary for the final product to comply with standards.

Purification of the Reaction Products

The mixture of fatty acids methyl esters (FAME) obtained from the transesterification reaction must be purified in order to comply with established quality standards for biodiesel. Therefore, FAME must be washed, neutralized and dried.

Successive washing steps with water remove the remains of methanol, catalyst and glycerin, since these contaminants are water-soluble. Care must be taken to avoid the formation of emulsions during the washing steps, since they would reduce the efficiency of the process. The first washing step is carried out with acidified water, to neutralize the mixture of esters. Then, two additional washing steps are made with water only. Finally the traces of water must be eliminated by a drying step. After drying, the purified product is ready for characterization as biodiesel according to international standards.

An alternative to the purification process described above is the use of ion exchange resins or silicates.

Glycerin as obtained from the chemical reaction is not of high quality and has no commercial value. Therefore, it must be purified after the phase separation. This is not economically viable in small scale production, due to the small glycerine yield. However, purification is a very interesting alternative for large-scale production plants, since, in addition to the high quality glycerin, part of the methanol is recovered for reutilization in the transesterification reaction (both from FAME and glycerin), and thus lowering biodiesel production costs. The steady increase of biodiesel production is fostering research for novel uses of glycerin in the production of high-value-added products.

It must be noted that the stages of the biodiesel production process are the same for all the production scales (laboratory, pilot plant small-, medium-, and large-scale industrial). However, the necessary equipment will be significantly different.

Production of Bioethanol

Ethanol can be produced from biomass by the hydrolysis and sugar fermentation processes. Biomass wastes contain a complex mixture of carbohydrate polymers from the plant cell walls known as cellulose, hemi cellulose and lignin. In order to produce sugars from the biomass, the biomass is pre-treated with acids or enzymes in order to reduce the size of the feedstock and to open up the plant structure. The cellulose and the hemi cellulose portions are broken down (hydrolysed) by enzymes or dilute acids into sucrose sugar that is then fermented into ethanol. The lignin which is also present in the biomass is normally used as a fuel for the ethanol production plants boilers. There are three principle methods of extracting sugars from biomass. These are concentrated acid hydrolysis, dilute acid hydrolysis and enzymatic hydrolysis.

Concentrated Acid Hydrolysis Process

The Arkanol process works by adding 70-77% sulphuric acid to the biomass that has been dried to a 10% moisture content. The acid is added in the ratio of 1.25 acid to 1 biomass and the temperature is controlled to 50C. Water is then added to dilute the acid to 20-30% and the mixture is again heated to 100C for 1 hour. The gel produced from this mixture is then pressed to release an acid sugar mixture and a chromatographic column is used to separate the acid and sugar mixture.

Dilute Acid Hydrolysis

The dilute acid hydrolysis process is one of the oldest, simplest and most efficient methods of producing ethanol from biomass. Dilute acid is used to hydrolyse the biomass to sucrose. The first stage uses 0.7% sulphuric acid at 190C to hydrolyse the hemi cellulose present in the biomass. The second stage is optimised to yield the more resistant cellulose fraction. This is achieved by using 0.4% sulphuric acid at 215C.The liquid hydrolates are then neutralised and recovered from the process.

Enzymatic Hydrolysis

Instead of using acid to hydrolyse the biomass into sucrose, we can use enzymes to break down the biomass in a similar way. However this process is very expensive and is still in its early stages of development.

Wet Milling Processes

Corn can be processed into ethanol by either the dry milling or the wet milling process. In the wet milling process, the corn kernel is steeped in warm water, this helps to break down the proteins and release the starch present in the corn and helps to soften the kernel for the milling process. The corn is then milled to produce germ, fibre and starch products. The germ is extracted to produce corn oil and the starch fraction undergoes centrifugation and saccharifcation to produce gluten wet cake. The ethanol is then extracted by the distillation process. The wet milling process is normally used in factories producing several hundred million gallons of ethanol every Year.

Dry Milling Process

The dry milling process involves cleaning and breaking down the corn kernel into fine particles using a hammer mill process. This creates a powder with a course flour type consistency. The powder contains the corn germ, starch and fibre. In order to produce a sugar solution the mixture is then hydrolysed or broken down into sucrose sugars using enzymes or a dilute acid. The mixture is then cooled and yeast is added in order to ferment the mixture into ethanol. The dry milling process is normally used in factories producing less than 50 million gallons of ethanol every Year.

Sugar Fermentation Process

The hydrolysis process breaks down the cellulostic part of the biomass or corn into sugar solutions that can then be fermented into ethanol. Yeast is added to the solution, which is then heated. The yeast contains an enzyme called invertase, which acts as a catalyst and helps to convert the sucrose sugars into glucose and fructose (both $C_6H_{12}O_6$).

The chemical reaction is shown below:

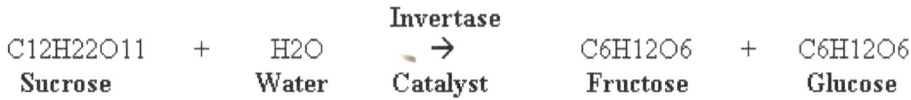

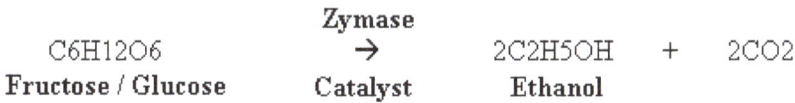

$$C12H22O11 \ + \ H2O \ \xrightarrow{\text{Invertase}} \ C6H12O6 \ + \ C6H12O6$$

| Sucrose | Water | Catalyst | Fructose | Glucose |

The fructose and glucose sugars then react with another enzyme called zymase, which is also contained in the yeast to produce ethanol and carbon dioxide.

The chemical reaction is shown below:

$$C6H12O6 \ \xrightarrow{\text{Zymase}} \ 2C2H5OH \ + \ 2CO2$$

Fructose / Glucose   Catalyst   Ethanol

The fermentation process takes around three days to complete and is carried out at a temperature of between 250C and 300C.

Fractional Distillation Process

The ethanol, which is produced from the fermentation process, still contains a significant quantity of water, which must be removed. This is achieved by using the fractional distillation process. The distillation process works by boiling the water and ethanol mixture. Since ethanol has a lower boiling point (78.3C) compared to that of water (100C), the ethanol turns into the vapour state before the water and can be condensed and separated.

Bioethanol Production by Algae

Algae as Bioethanol Feedstock

Algae as bioethanol feedstock Algae are simple organisms containing chlorophyll and they use light for photosynthesis. Algae can grow phototrophically or heterotrophically. Phototrophic algae convert carbondioxide in atmosphere to nutrients such as carbohydrate. Conversely, heterotrophic algae continue their development by utilizing organic carbon sources. Algae can grow in every season and everywhere such as salty waters, fresh waters, lakes, deserts and marginal fields etc. However for their cultivation, generally open systems like ponds and photobioreactors as closed systems are used. Open ponds are the most used cultivation systems in industry. They are more preferable than other systems due to having low investment and operation costs. On the other hand difficult control of cultivation conditions and contamination risk are the main disadvantages of the open systems. Besides being cheap and low energy need, their cleaning also can be done easily. Although, open tanks have low cost and easy operation, parameters like light intensity, temperature, pH and dissolved oxygen concentration cannot be controlled easily. Most produced algae species in open systems are Spirulina, Chlorella and Dunaliella. In comparison with open systems, photobioreactors have very high photosynthetic efficiency. Thus, photobioreactors ensures high biomass yield. Though they are expensive, they are preferred for specific algal production. Algal production which is controlled in terms of parameters like light, pH, carbon dioxide etc., can be achieved and also contamination risk is not seen mostly in photobioreactors. Since they are closed systems, evaporation doesn't occurred and they enable production of special biochemical materials. Although there are many types of photobioreactors, most commonly systems are vertical and horizontal tubular columns and flat-type photobioreactors. These photobioreactors which are made of glass or plastic, can be designed in shapes of horizontal vertical, conical or curved etc.

Algae are classified as microalgae and macroalgae. Microalgae as their name implies, are prokaryotic or eukaryotic photosynthetic microorganisms. They can survive in hard conditions with their unicellular or simple colony structures. Because of being photosynthetic organism, they can produce high amount of lipid, protein and carbohydrate in a short time. Besides biodiesel and bioethanol there are lots of high value products and sub-products produced from microalgae such as biogas, biobutanol, acetone, Omega 3 oil, eicosapentaenoic acid, livestock feed, pharmaceuticals and cosmetics. Especially sub-products are preferred for economic support of main process. Chemical composition of microalgae can change according to the cultivation type and cultivation conditions. They can have rich or balanced composition of protein, lipid and carbohydrate amounts. Microalgae especially get attention due to have high lipid content. Many species of microalgae accumulate a significant amount of lipids in their structure and can provide high oil yield. Their average lipid content can change between 1-70%, but this ratio can reach up to 90% of dry weight under certain conditions. Macroalgae or seaweed are plants which are adapted to the marine life, often located in coastal areas. They are classified as brown seaweeds, red seaweeds and green seaweeds according to their pigments. Due to have high photosynthesis capability, they have sufficient carbon source for usage in biorefinery. On the contrary of their appearances, their features of morphologic and physiological and chemical compositions are different from terrestrial plants. Unlike the structure of the lignocellulosic biomass of microalgae, they comprise substances such as carrageenan, laminaran, mannitol, alginate which are used in various sectors. They are separated from microalgae with having low lipid content and different from lignocellulosic material with having less or no lignin in their structure.

Microalgae stand out as biodiesel feedstock with the ability of lipid production and high photosynthetic efficiency. As for macroalgae, they are utilized for biogas or bioethanol production with their carbohydrates. First studies as algal biofuels are focused on biodiesel production. However, there is a potential for carbohydrates in the structure of algae which can be utilized for ethanol production after various hydrolysis processes. Algal cells in the water don't need structural biopolymers such as hemicellulose and lignin which are necessary for terrestrial plants. This simplifies the process of bioethanol production. Marine algae can produce high amount of carbohydrate every year. Also it is expected that algae will meet the demand of biofuel feedstock due to harvest in a short time than other biofuel raw materials. Microalgae which have high amount of starch such as Chlorella, Dunaliella, Chlamydomonas, Scenedesmus are very useful for bioethanol production. In addition to that, microalgae don't need energy consumption for distribution and transportation of molecules like starch. Like microalgae, macroalgae are also raw materials that can be used in ethanol production. Absence of lignin or having less lignin in the structure, simplifies the hydrolysis stages. Although it changes with the algal species, they have various amounts of heteropolysaccharides in their structures. Whereas red algae contain carrageenan and agar, brown algae have laminaran and mannitol in their structure.

Algal Polysaccharides

Ethanol production from algae is based on fermentation of algal polysaccharides which are starch, sugar and cellulose. For microalgae, their carbohydrate content (mostly starch) can be reached to 70% under specific conditions. Microalgal cell walls are divided into inner cell wall layer and outer cell wall layer. Outer cell layer can be trilaminar outer layer and thin outer monolayer. Also there can be no outer layers as well. Outer cell walls of microalgae contain certain polysaccharides such as pectin, agar and alginate. However their composition can be vary from species to species. On the contrary, inner cell walls of microalgae constitute mostly cellulose, hemicellulose and other materials. Due to have cellulose in their cell walls and starch, microalgae are considered as a feedstock for production of bioethanol. Most of their cell wall polysaccharides and starch can be fermented for bioethanol production.

Similarly, carbohydrate content of macroalgae is found 25-50% in the green algae, 30-60% in the red algae and 30-50% in the brown algae. Macroalgae species which have the highest polysaccharide content are Ascophyllum (42-70%), Porphyra (40-76%) and Palmaria (38-74%). High carbohydrate content of algal species are presented in table below. Polysaccharides in the cell wall of macroalgae are composed of cellulose and hemicelluloses. Cellulose and hemicelluloses content of macroalgae compose 2-10% of dry weight. Lignin is only exists in Ulva species and it constitutes 3% of dry weight. Differently from microalgae, alginate, mannitol, glucan and laminarin are the most abundant polymers in macroalgal structure. Alginates are

Table: Carbohydrate content of algal species

Algal species	Carbohydrate content (%)
C. vulgaris	55.0
Chlamydomonas reinhardtii	UTEX 90 60.0
Chlorococum sp.	32.5
S. obliquus CNW-N	51.8

Tetraselmis sp.CS-362	26.0
Ulva lactuca	55-60
Ascophyllum	42-70
Porphyra	40-76
Palmaria	38-74

polymers which are obtained from cell walls of various brown algae. They consist of mannuronic acid and L-gluronic acid monomers and they are extracted from cell walls by using sodium carbonate. Although they are usually used as stabilizer in pharmaceutical industry, they also used in paper and adhesive manufacture, oil, photography and textile industries. Caragenan is another polysaccharide which is obtained from red algae. It is used as stabilizer in food, textile and pharmaceutical industry. Agar is also acquired from red algae. Like caragenans and alginates, it is extracted with hot water and used as stabilizer and gelling agent. 90% of produced agar is utilized in the food industry, the remaining 10% is used in microbiological and biotechnological field. Mannitol which is a structure in brown algae is a sugar alcohol, especially found in Laminaria and Ecklonia. Mannitol content of macroalgae can change with season and environmental conditions. Mannitol can be used in pharmaceutical, paint, leather and paper manufacture. In addition to that, mannitol can be utilized in food industry as a coating material. Laminarin is a polysaccharide which helps the immune system by increasing the B cells, provides protection against infection by bacterial pathogens and severe irradiation. Another polysaccharide from macroalgae is ulvan. It is mainly presented in Ulva sp. and it is source of sugars for production of fine chemicals.

Pre-treatment Technologies for Biomass

The most important difficulties encountered in the production of bioethanol are the pretreatment of biomass. The objectives of an effective pre-treatment are obtaining sugars directly or later by hydrolysis, preventing lost or degradation of obtained sugars, limiting the toxic materials which inhibit the ethanol production, reducing energy requirement for process and minimizing the production cost. There are four pre-treatment techniques including physical, chemical, physicochemical and biological pre-treatments that are applied to biomass. Pre-treatment process is the step that forms the significant part of the cost of ethanol production. Although there is no technique that can be considered as the best option, researches and developments are carried on to reduce cost and improve performance.

Physical Pre-treatments

Mechanical Comminution

Chipping, grinding and milling are the most used techniques for mechanical comminution.

Comminutions improve the efficiency of the process for the next steps by reducing the polymerization degree and increase the specific surface by reducing cellulose cristallinity.

Energy that is need in the process depends on the initial and final dimensions of particles, moisture content and structure of the raw material. In order to assist enzymatic hydrolysis of lignocellulosic materials various milling techniques can be used. For instance, pretreatment of rice straw with wet disk milling gave higher hydrolysis yields than usual dry milling.

Pyrolysis

Pyrolysis is an endothermic process which is a reaction needs low energy input and treats biomass over the temperature of 300°C and degrades cellulose to char and gaseous products like CO and H_2. When the char is washed with water or diluted acid, remaining solution contains sufficient amount of carbon source to support microbial growth for the production of bioethanol. Approximately 55% of biomass weight is lost in the washing step. It is reported in a study that Fan et al. have performed 80-85% conversion of cellulose to reducing sugars.

Microwave Oven Pre-treatment

Microwave oven pre-treatment is a simple method with short reaction time, high heating efficiency and low energy input. Thermal and non-thermal effects which are generated by microwaves in a liquid medium are used in this technique. The heat generated in biomass results in a polar bond vibration. This causes an explosion between the particles and degradation of lignocellulosic structure. Asetic acid is released from lignocellulosic material and an acidic medium is occurred for hydrolysis. Ooshima et al. investigated the effect of microwave pre-treatment on rice straw and baggase and it was found that an improvement in total reducing sugar production. In recent years, microwave pre-treatments are carried out with various chemical reagents and their potential are investigated. In the studies of alkali microwave pre-treatment, NaOH provides higher reducing sugar yields on switchgrass and coastal bermudagrass in comparison with other alkaline reagents such as Na_2CO_3, $Ca(OH)_2$. Also for pre-treatment of rice straw and its hulls, this technique made cellulose more accessible to enzymes.

Physicochemical pre-treatments

Steam Explosion Method

Steam explosion method is a technique that provides accessibility on the biomass for degradation of cellulose. This method comprise the heating of biomass under high pressure steam (20–50 bar, 160 270 °C) for a few minutes, then reaction is stopped when the pressure conditions arrive to the atmospheric conditions. Diffusion of the steam into the lignocellulosic matrix leads to the dispersion of fibers. No catalyst is used during the applied method. Levulinic acid, xylitol and alcohols are obtained after the degradation of biomass. Many types of biomass such as poplar, eucalyptus, olive residues, corn stover, wheat straw, sugarcane bagasse, grasses have been pre-treated with steam explosion method efficiently.

Liquid hot Water Method

Liquid hot water method treats biomassby using water which is kept in a liquid state under high pressure and temperature for 15 minutes without adding any chemical or catalyst. Instead of steam explosion method, this technique does not need rapid pressure drop or expansion. Pressure is used to prevent evaporation and to stabilize the water in this method. Although it provides the release of hemisellulosic sugars as oligomers, it causes the formation of little amounts of undesirable components which inhibit microbial growth such as carboxylic acid, furfural. Since there is no need for chemicals, it is an environmental and economic method. It is reported that liquid hot water method improves the enzymatic hydrolysis by removing 80% of hemicelluloses when it is pre-treated corn stover, sugarcane bagasse and wheat straw.

Ammonia Ffiber Explosion (AFEX)

Ammonia fiber explosion (AFEX) is a method that liquid ammonia and steam explosion are carried out together. In this method, biomass which has 15-30% moisture content is treated with liquid ammonia at a loading ratio of 1–2 kg NH3/kg dry biomass. To acquire appropriate temperature, pressure over 12 atm is required. Whereas being an easy method and have short reaction time, it is not effective on raw materials that contain high lignin content. Ammonia has effects such as shredding biomass fibers, partially decrystallization of cellulose and destroying carbohydrate attachments. Although sugars are not released directly with this method, it enhances polymers (hemicellulose and cellulose) to be attacked enzymatically. Thus, low amount of enzyme is enough for enzymatic hydrolysis after AFEX. In order to improve the process economically, ammonia must be recover after the pre-treatment. Ammonia loading, temperature, high pressure, moisture content of biomass, and residence time are the basic parameters which effect AFEX process. Up to 90% cellulose and hemicelluloses conversions can be acquired with this technique.

CO_2 explosion

CO_2 explosion is similar to AFEX method. However this method has low process cost due to need low temperature. Also formation of inhibitors in the steam explosion is not occurred in this technique. In addition to that, its conversion yields are very high compared to steam explosion.

Wet Oxidation

Wet oxidation method is based on the treatment of biomass with water and air or oxygen as a catalyst over the temperature of 120 °C. Although solubility of hemicellulose and lignin are increased with this method, free hemicelluloses molecules do not hydrolyze. Whereas sugar monomers are formed in steam explosion and dilute acid pre-treatment, sugar which released in wet oxidation method are oligomers. In a study performed by Pederson et al.40% glucose yield was obtained for wet oxidation of wheat straw.

Chemical Pre-treatments

Chemical pre-treatments include dilute acid, alkaline, ammonia, organic solvent pre treatments and methods that use other chemicals. These processes are easy to perform and also good conversion yields are achieved in a short time.

Acid Pre-treatment

Acid pre-treatments are methods that acid is used as catalyst to make cellulose more accessible to the enzymes. These processes are divided into two groups as using concentrated acid or diluted acid. Using concentrated acid is less preferable than dilute acid because of forming high amount of inhibiting components and causing corrosion in the equipments. Generally sulphuric acid, hydrochloric acid, nitric acid and phosphoric acid are used in these pretreatments. Dilute acid are applied at moderate temperatures to convert lignocellulosic structures to soluble sugars. Nowadays biomass is pre-treated with dilute sulphuric mostly to hydrolyze hemicelluloses and facilitate enzymatic hydrolysis. Dilute sulphuric acid hydrolyzes biomass to hemicelluloses, and then hydrolyzes to xylose and other sugar and break xylose down to furfural. Furfural which is a toxic component in ethanol production process, is recovered by distillation. Miranda et al. have investigated the effect of acid pretreatments with the concentrations between 0.05-10 N, and have obtained the highest sugar yield under the condition of 2 N acid pre-treatment. In their experiments, 2 N to 10 N acid pretreatments, it is reported that a decrease have been observed in sugar yields. Larsson et al. also mentioned that in an experiment about acid pre-treatment of soft wood, a decrease in ethanol yields have been observed with an increasing acid concentration. In addition to this, it is indicated that formic acid which is a toxic molecule, is presented in the media and inhibits the fermentation.

Alkaline Pre-treatment

These processes are carried out at low temperature and pressure compared to other techniques. Unlike acid pre-treatments, lignin can be removed without major effects on the other components. However there are limitations such as transformation of some alkaline to unrecoverable salts. In addition to that, solubility of hemicelluloses and cellulose are less in this pre-treatment compared to solubility in acid pre-treatment. Alkaline pre-treatment reduces the lignin and hemicelluloses content of biomass and improves the surface area and helps water molecules for breaking bonds between hemicelluloses and lignin. The most used catalysts in this method are sodium hydroxide, potassium hydroxide, calcium hydroxide and ammonia. Effects of alkaline pre-treatments are varies according to biomass. In an alkaline pretreatment of coastal bermudagrass, reducing sugar yields are decrease with an increasing alkaline concentration. However, Wang et al. reported that under the conditions of increasing alkaline concentrations, glucose yields were increased. Like dilute acid pretreatments, dilute alkaline pre-treatments also can form inhibitory by-products such as furfural, hydroxymethylfurfural and formic acid.

Organosolv Pre-treatment

Organosolv pre-treatment is a process that uses organic solvents such as methanol, ethanol, acetone, ethylene glycol. Catalysts are also can be added to the process along with solvents.

Hydrochloric acid, sulphuric acid, sodium hydroxide and ammonia are the catalysts used in the process. Besides bonds of lignin and hemicellulose can be broken, pure and high quality lignin can be obtained as a by-product. Removal of lignin improves the surface area and provides accessibility of enzymes to cellulose. After the pre-treatment, cellulosic fibers, solid lignin and liquid solution of hemicellulose sugars are obtained. This method has some disadvantages like oxidation, volatilization and creating high risk in process at high pressure.

Also solvents must be recovered due to formation of significant amounts of furfural and soluble phenols and to reduce operation cost.

Biological Pre-treatments

Compared to the above methods applied to the production of bioethanol, using fungi in pretreatments is considered environmentally friendly because of not using chemicals, less energy input, not required reactors that resistant to corrosion and pressure, and minimum inhibitor formation. Fungi which are used in biological pre-treatments are generally brown, white and soft mold. These fungi can be degrade lignin, hemicelluloses and cellulose partially. Despite of its advantages, long process time, large production are and need of control continuously for growth of microorganisms ensue as disadvantages for commercial productions.

Enzymatic hydrolysis is the step of hydrolysis of cellulose by specific cellulase enzymes. Obtained products after hydrolysis are reducing sugars that include glucose. Cost of the enzymatic hydrolysis are less than acid or alkaline hydrolysis due to reaction is carried out under mild conditions (4.8 pH, temperature of 45-50 °C). Cellulase enzymes that are used in hydrolysis can be produced by bacteria and fungi. These microorganisms can be aerobic, anaerobic, mesofilic or thermophilic. Bacteria which produce cellulase can be exemplify as Clostridium, Cellulomonas, Bacillus, Thermomonospora, Ruminococcus, Bacteriodes, Erwinia, Acetovibrio, Microbispora and Streptomyces. Trichoderma, Penicillium, Fusarium, Phanerochaete, Humicola and Schizophillum sp. are identified as cellulase produced fungi among the fungi. Although there are anaerobic bacteria which produce cellulase with high specific activity, these bacteria are not suitable for commercial productions. Cellulase enzymes consist of mixture of endoglucanase, exoglucanase and b-glucosidase. While endoglucanase attacks the regions where cellulose fibers have low crystallinity, exoglucanase removes the cellulose units from released chains with the effect of endoglucanase and then degrades the molecule. B-glucosi-dase hydrolyzes the cellulose units and enables the formation of glucose. Enzymatic hydrolysis can be affected by certain factors which are enzyme-related and substrate-related factors. Substrate-related factors have a directly influence on enzymatic hydrolysis. These factors are connected to each other and effect the enzymatic conversion. These factors can be defined as degree of polymerization and crystallinity of cellulose, accessibility of the substrate, lignin and hemicelluloses content and pore size.

Hydrolysis rates of biomass depend on the degree of polymerization and crystallinity of cellulose. Degree of polymerization is related to crystallinity. Cellulase enzymes can hydrolyze the crystalline structure of cellulose. Endoglucanase enzymes decrease polymerization degree of cellulosic component by cutting the internal sites of cellulose chains in the enzymatic hydrolysis. Accessibility of the substrate is another main factor effect hydrolysis rate. The rate of hydrolysis increases with increasing substrate accessibility because of being surface area more available for enzymatic attack. Lignin and hemicellulose are complex structures to hydrolyze in lignocellulosic materials. Due to have a role like cement, lignin acts as physical barrier and prevents the digestible parts of cellulose to hydrolyze and it becomes very difficult for enzymes to access cellulose. For this reason, they reduce the efficiency of hydrolysis. Removal of hemicellulose enhances the pore size and provides accessibility to cellulose for enzymes in order to perform hydrolysis efficiently. Pore size of the substrate is one of the limiting factors in enzymatic hydrolysis process. In many lignocellulosic material, external area of the biomass is smaller than internal area and this situation causes cellulase enzymes to entrap in the pores of the material. In order to increase hydrolysis rate, porosity of the biomass should be increased.

Fermentation

Fermentation is a process that based on disciplines of chemistry, biochemistry and microbiology and which fermentable sugars are converted to ethanol by microorganisms. Process consists of conversion of glucose to alcohol and carbon dioxide:

$$C_6H_{12}O_6 \rightarrow 2C_2H_5OH + 2CO_2$$

In this process 0.51 kg bioethanol and 0.49 kg carbon dioxide are obtained from per kg of glucose in theory maximum yield. However practically, microorganisms also use glucose for their growth, the actual yield is less than 100%. Microorganisms used in fermentation are utilized from 6-carbon sugars in ethanol production. Therefore, cellulosic biomass which have high amount of glucose are the materials that have easiest conversion capability. One of the most effective yeast which produces bioethanol is Saccharomyces cerevisiae. Besides having high bioethanol production yields, it has a resistance to high bioethanol concentration and inhibitor components which can be occurred after acid hydrolization of lignocellulosic biomass. Because reaction occurs under anaerobic conditions, oxygen molecules must be removed with nitrogen gas as a swept gas. Yeast and fungi can tolerate 3.5-5.0 pH ranges. S.cerevisiae has high osmotic resistance and can tolerate low pH levels like 4.0. Zymomonas stands out with rapid bioethanol production and high productivity compared to other traditional yeasts. However Z.mobilis cannot tolerate the toxic effects of asetic acid and various phenolic compounds in the lignocellulosic hydrolysate. Bioethanol yields of microorganisms are depend on temperature, pH level, alcohol tolerance, osmotic tolerance, resistance for inhibitors, growth rate and genetic stability. Fermentation processes generally are carried out with two basic processes as simultaneous saccharification and fermentation and separate hydrolysis and fermentation, yet new production processes have been developed.

Separate Hydrolysis and Fermentation (SHF)

Enzymatic hydrolysis is performed separately from fermentation in this process. Liquid which comes from hydrolysis reactor first converted to ethanol in a reactor that glucose fermented in, and then ethanol is distilled and remained unconverted ksilose is converted to ethanol in a second reactor. Advantage of the process is performing reactions in optimum conditions. On the other hand, usage of different reactors is increasing the cost. Also glucose and cellulose units that obtained after hydrolysis, inhibit activity of the enzyme and decrease hydrolysis rate.

Simultaneous Saccharification and Fermentation (SSF)

In this process, pre-treatment and enzymatic hydrolysis steps are carried out with fermentation step in the same reactor. It is very efficient when dilute acid or hot water at high temperature is applied in the process. High bioethanol yields can be achieved with SSF process. Also inhibiton of enzyme activity is very low due to fermenting glucose and cellulose units in the same media by yeast. Therefore, this process needs low amount of enzyme. In addition to that, process cost is reduced because of the reactions are carried out in one reactor. As a disadvantage, temperatures differences between saccharification and fermentation cause various effects in growth of microorganisms. Saccharomyces cultures are used in pH of 4.5 and temperature of 37 °C this process.

Simultaneous Saccharification and Co-Fermentation (SSCF) & Separate Hydrolysis and Co-Fermentation (SHCF).

Saccharomyces cerevisiae which used in fermentation cannot convert carbohydrates like pentos under moderate conditions and this causes impurity for biomass and decreases bioethanol yield. In order to overcome this, recombinant yeasts can be used to convert residues such as pentose to ethanol. In SSCF, recombinant yeasts and cellulase enzyme complex are fed to the same vessel to convert biomass to ethanol. This system is generally the same as SSF process. SCHF process is a combination of SSCF and SHF. In this process, fermentation and hydrolysis are carried out in different vessel. This process can produce ethanol with high productivity in comparison with SHF process.

Distillation and Purification

A distillation process is necessary for separation of ethanol from mixture and purification of ethanol after fermentation process. Process is performed simply with boiling ethanol-water mixture. Because of boiling point of water (100°C) is higher than boiling point of ethanol (78 °C), ethanol vaporized before water. However, due to being an azeotrop mixture, high amount of energy is used for distillation. In order to separate azeotrop mixtures, an agent which changes the azeotrop structure must be added to the mixture. Added substance changes the volatility of mixture by effecting the molecular attractions in the mixture. Various separation agents such as benzene, pentane, cyclohexane, hexane, acetone, and diethyl ether can be used in this process. Distillation column which has two streams as top and bottom, separates most of the bioethanol from the mixture. While top stream is rich in bioethanol, bottom stream is rich in water. 37% bioethanol then concentrated in rectifying column to approach concentration of 95%. Product which is remained in the bottom is fed to stripping column in order to remove excess water. Mostly in plants, recovery of bioethanol in distillation columns is fixed to be 99.6% due to decrease bioethanol loss.

Production of Biogas (bio-methane)

Absorption

The separation principle of absorption is based on differing solubilities of various gas components in a liquid scrubbing solution. In an upgrading plant using this technique the raw biogas is forced into intensive contact with a liquid. A scrubbing column is filled with a plastic packing in order to increase the contact area between the phases. The components to be removed from the biogas (mostly carbon dioxide) are typically far more soluble in the applied liquid than methane and are removed from the gas stream. As a result, the remaining gas stream is enriched with methane and the scrubbing liquid leaving the column is rich in carbon dioxide. In order to maintain the absorption performance, this scrubbing liquid has to be replaced by fresh liquid or regenerated in a separate step (desorption or regeneration step). Currently, three different upgrading technologies embodying this physical principle are available.

Physical Absorption: Pressurised Water Scrubbing

The absorbed gas components are physically bound to the scrubbing liquid, in this case water. Carbon dioxide has a higher solubility in water than methane and will therefore be dissolved to a higher extent, particularly at lower temperatures and higher pressures. In addition to carbon dioxide, hydrogen sulphide and ammonia can also be reduced in the biomethane stream using water as a scrubbing liquid. The effluent water leaving the column is saturated with carbon dioxide and is transferred to a flash tank where the pressure is abruptly reduced and the major share of the dissolved gas is released. As this gas mainly contains carbon dioxide, but also a certain amount of methane (methane is also soluble in water, but to a smaller extent) this gas is piped to the raw biogas inlet. If the water is to be recycled back to the absorption column, it has to be regenerated and is therefore pumped to a desorption column where it meets a counter current flow of stripping air, into which the remaining dissolved carbon dioxide is released. The regenerated water is then pumped back to the absorber as fresh scrubbing liquid.

The drawback of this method is that the air components – oxygen and nitrogen – are dissolved in the water during regeneration, and thus are transported to the upgraded biomethane gas stream. Therefore, biomethane produced with this technology always contains oxygen and nitrogen. As the produced biomethane stream is also saturated with water, the final upgrading step typically is gas drying, for example by the application of glycol scrubbing.

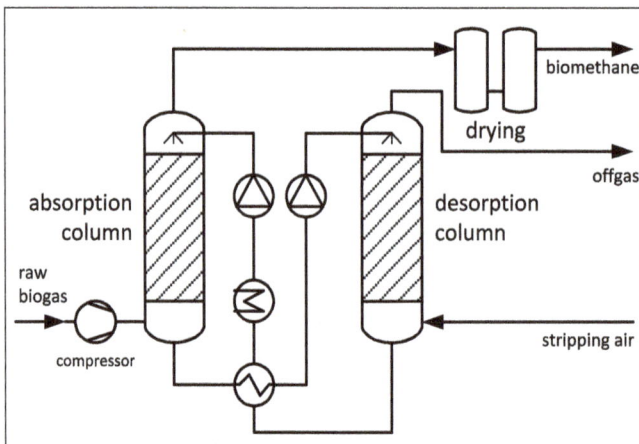

Flowsheet of a typical biogas upgrading unit applying pressurised water scrubbing; picture of the upgrading plant Könnern, Germany with a raw biogas capacity of 1,250m³/h

Organic Physical Absorption

Using a basic similar concept to water scrubbing, this technology uses an organic solvent solution (e.g. polyethylene glycol) instead of water as a scrubbing liquid. Carbon dioxide shows higher solub-ilities in these solvents than in water. As a result, less scrubbing liquid circulation and smaller apparatus are needed for the same raw biogas capacity.

Examples of commercially available biogasupgrading technologies implementing organic physical scrubbing are Genosorbâ, Selexolâ, Sepasolvâ, Rektisolâ and Purisolâ.

Chemical Absorption: Amine Scrubbing

Chemical absorption is characterised by a physical absorption of the gaseous components in a scrubbing liquid, followed by a chemical reaction between scrubbing liquid components and ab-sorbed gas components within the liquid phase. As a result, the bonding of unwanted gas comp-onents to the scrubbing liquid is significantly stronger, and the loading capacity of the scrubbing liquid is several times higher. The chemical reaction is strongly selective and the amount of methane also absorbed in the liquid is very low, resulting in very high methane recovery and very low methane slip. Due to the high affinity of especially carbon dioxide to the solvents used (mainly aqueous solutions of Monoethanolamine MEA, Diethanolamine DEA and Methyldiethanolamine MDEA), the operating pressure of amine scrubbers can be kept significantly lower compared to pressurised water scrubbing plants of similar capacity.

Typically, amine scrubbing plants are operated at the slightly elevated pressure already arising in the raw biogas, and no further compression is needed. The high capacity and high selectivity of the amine solution, although an advantage during absorption, turns out to be a disadvantage during the regeneration of the scrubbing solution. Chemical scrubbing liquids require a significantly increased amount of energy during regeneration which has to be provided as process heat. The loaded amine solution is heated up to about 160°C where most of the carbon dioxide is released and leaves the regeneration column as a considerably pure offgas stream. As a small part of the scrubbing liquid is lost to the produced biomethane due to evaporation, it has to be replenished frequently. Hydrogen sulphide could also be absorbed from the raw biogas by chemical absorption, but even higher temperatures during regeneration would be needed. That is why it is advisable to remove this component prior to the amine scrubber.

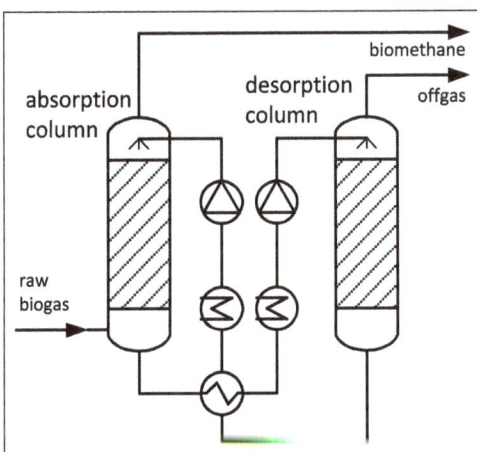

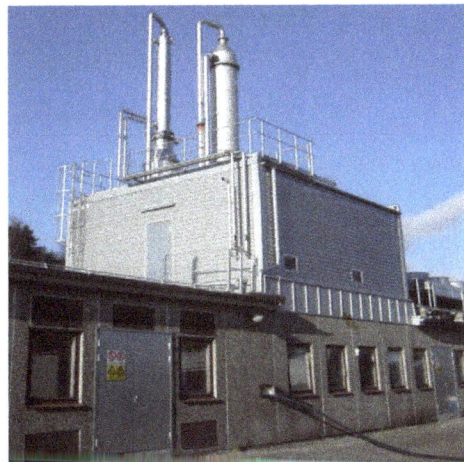

Flowsheet of a typical biogas upgrading unit applying amine scrubbing; picture of the upgrading plant Gothenburg, Sweden with a raw biogas capacity of 1,600m³/h

Adsorption: Pressure Swing Adsorption (PSA)

Gas separation using adsorption is based on different adsorption behaviour of various gas components on a solid surface under elevated pressure. Usually, different types of activated carbon or molecular sieves (zeolites) are used as the adsorbing material. These materials selectively adsorb carbon dioxide from the raw biogas, thus enriching the methane content of the gas. After the adsorption at high pressure the loaded adsorbent material is regenerated by a stepwise decrease in pressure and flushing with raw biogas or biomethane. During this step offgas is leaving the adsorber. Afterwards, the pressure is increased again with raw biogas or biomethane and the adsorber is ready for the next sequence of loading. Industrial scale upgrading plants implement four, six or nine adsorber vessels in parallel, at different positions within this sequence in order to provide a continuous operation. During the decompression phase of the regeneration, the composition of the offgas is changing as the co-adsorbed methane is released earlier (at higher pressures) and the bulk of carbon dioxide is preferentially desorbed at lower pressures. Thus, the offgas from the first decompression stages is typically piped back to the raw biogas inlet in order to reduce the methane slip. Offgas from later stages of regeneration could be led to a second stage of adsorption, to the offgas treatment unit or could be vented to the atmosphere. As water and hydrogen sulphide contents in the gas irreversibly harm the adsorbent material, these components have to be removed before the adsorption column.

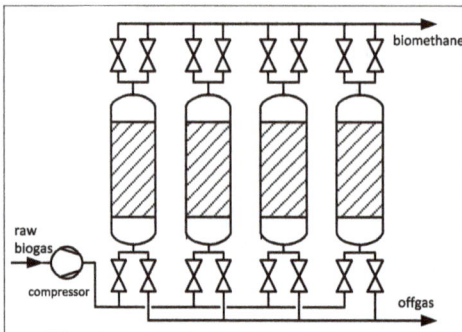

Flowsheet of a typical biogas upgrading unit applying pressure swing adsorption; picture of the upgrading plant Mühlacker, Germany with a raw biogas capacity of 1,000m³/h

Membrane Technology: Gaspermeation

Membranes for biogas upgrading are made of materials that are permeable for carbon dioxide, water and ammonia. Hydrogen sulphide, oxygen and nitrogen permeate through the membrane to a certain extent and methane passes only to a very low extent. Typical membranes for biogas upgrading are made of polymeric materials like polysulfone, polyimide or polydimethylsiloxane. These materials show favourable selectivity for the methane/carbon dioxide separation combined with a reasonable robustness to trace components contained in typical raw biogases. To provide sufficient membrane surface area within compact plant dimensions, these membranes are applied in the form of hollow fibres, combined to assemble a number of parallel membrane modules.

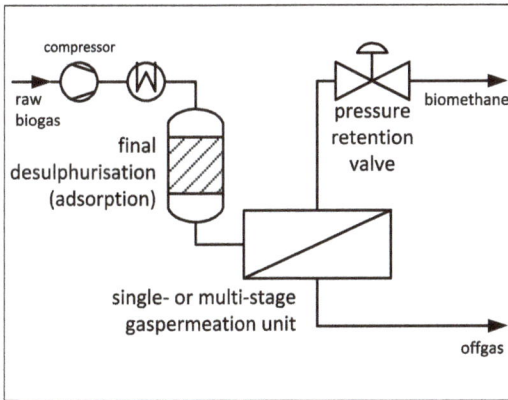

Flowsheet of a typical biogas upgrading unit applying the membrane technology gaspermeation; picture of the upgrading plant Kisslegg, Germany with a raw biogas capacity of 500m³/h

After the compression to the applied operating pressure the raw biogas is cooled down for drying and removal of ammonia. After reheating with compressor waste heat the remaining hydrogen sulphide is removed by means of adsorption on iron or zinc oxide. Finally, the gas is piped to a single- or multi-staged gas permeation unit. The numbers and interconnection of the applied membrane stages are not determined by the desired biomethane quality, but by the requested methane recovery and specific compression energy demand. Modern upgrading plants with more complex design offer the possibility of very high methane recoveries and relatively low energy demand. Even multi-compressor arrangements have been realised and proved to be economically advantageous. The operation pressure and compressor speed are both controlled to provide the desired quality and quantity of the produced biomethane stream.

Comparison of Different Biogas Upgrading Technologies

It is hard to give a universally valid comparison of the different biogas upgrading technologies because many essential parameters strongly depend on local circumstances. Furthermore, the technical possibilities of a certain technology (for example regarding the achievable biomethane quality) often do not correspond with the most economic operation. The technical development maturity of most biogas upgrading methods nowadays is typically sufficient to meet any needs of a potential plant operator. It is mainly a question of finding a plant design providing the most economic operation for biomethane production. As a result, it is strongly recommended to perform a detailed analysis of the specific biomethane costs to be expected and to account for all possible upgrading technologies. As a guiding tool to fulfil these tasks the "Biomethane

Calculator" has been developed during this project and will be updated every year. This tool models all relevant upgrading steps and upgrading technologies and allows for a qualified estimation of specific biomethane production costs to be expected.

The following table summarises the most important parameters of the described biogas upgrading technologies, applied to a typical raw biogas composition. Values of certain parameters represent averages of realised upgrading plants or verified data from literature.

Membrane technology offers the possibility to widely adapt the plant layout to the local circum-stances by the application of different membrane configurations, multiple membrane stages and multiple compressor variations. This is why a certain range is given for most of the parameters. The first number always corresponds to the simpler plant layout ("cheaper" and with low methane recovery) while the other higher number corresponds to a high recovery plant layout.

Parameter	Water scrubbing	Organic physical scrubbing	Amine scrubbing	PSA	Membrane technology
typical methane content in biomethane [vol%]	95-99	95-99	>99	95-99	95-99
methane recovery [%]	98	96	99.96	98	80-99.5
methane slip [%]	2.0	4.0	0.04	2.0	20-0.5
typical delivery pressure [bar(g)]	4-8	4-8	0	4-7	4-7
electric energy demand [kWhel/m³ biomethane]	0.46	0.49-0.67	0.27	0.46	0.25-0.43
heating demand and temperature level	-	medium 70-80°C	high 120-160°C	-	-
desulphurisation requirements	process dependent	yes	yes	yes	yes
consumables demand	antifouling agent, drying agent	organic solvent (non-hazardous)	amine solution (hazardous, corrosive)	activated carbon (non- hazardous)	
partial load range [%]	50-100	50-100	50-100	85-115	50-105
number of reference plants	high	low	medium	high	low
typical investment costs [€/(m³/h) biomethane]					
for 100m³/h biomethane	10,100	9,500	9,500	10,400	7,300-7,600
for 250m³/h biomethane	5,500	5,000	5,000	5,400	4,700-4,900
for 500m³/h biomethane	3,500	3,500	3,500	3,700	3,500-3,700
typical operational costs [ct/m³ biomethane]					

for 100m³/h biomethane	14.0	13.8	14.4	12.8	10.8-15.8
for 250m³/h biomethane	10.3	10.2	12.0	10.1	7.7-11.6
for 500m³/h biomethane	9.1	9.0	11.2	9.2	6.5-10.1

Removal of Trace Components: Water, Ammonia, Siloxanes, Particulates

Biogas is saturated with water vapour when it leaves the digester. This water tends to condense within apparatus and pipelines and, together with sulphur oxides, may cause corrosion. By incr-easing the pressure and decreasing the temperature water will condense from the biogas and can thereby be removed. Cooling can either be realised using the ambient surroundings (air, soil) or by active cooling (i.e., refrigeration). Water can also be removed by scrubbing with glycol or by adsorption on silicates, activated charcoal or molecular sieves (zeolites).

Ammonia is usually separated when the biogas is dried by cooling, as its solubility in liquid water is high. Furthermore, most technologies for carbon dioxide removal are also selective for the removal of ammonia. A separate cleaning step is therefore usually not necessary.

Siloxanes are used in products such as deodorants and shampoos, and can therefore be found in biogas from sewage sludge treatment plants and landfill gas. These substances can create serious problems when burned in gas engines or combustion facilities. Siloxanes can either be removed by gas cooling, by adsorption on activated carbon, activated aluminium or silica gel, or by absorption in liquid mixtures of hydrocarbons.

Particulates and droplets can be present in biogas and landfill gas and can cause mechanical wear in gas engines, turbines and pipelines. Particulates that are present in the biogas are separated by fine mechanical filters (0.01μm – 1μm).

Removal of Methane from the Offgas

The offgas produced during biogas upgrading still contains a certain amount of methane depending on the methane recovery of the applied gas separation technology. As methane is a strong greenhouse gas, it is of vital importance for the overall sustainability of the biomethane production chain to minimise the methane emissions to the atmosphere. It should be noted that the emissions of methane from biogas processing plants is restricted in most countries. Additionally, higher amounts of methane in the offgas increase the specific upgrading costs and could inhibit economic plant operation. But it is never so simple, as there is a trade-off in selecting a certain methane recovery value because a higher methane recovery always increases investment and operational costs of a certain upgrading technology. As a result, the most promising plant layout in terms of economics usually accepts a certain amount of methane left in the offgas and applies a certain treatment of the gas prior to venting it to the atmosphere.

The most common technique for removing the methane content in the offgas is oxidation (i.e., combustion) and generation of heat. This heat can either be consumed at the anaerobic digestion plant itself (as this plant often has a heat demand), it can be fed to a district heating system (if locally available) or it has to be wasted by cooling. Another possibility would be to mix the offgas with raw biogas and feed it to an existing CHP gas engine. Either way, the layout of the plant has to be planned carefully, since the offgas of a modern biogas upgrading plant seldom contains enough methane to maintain a flame without addition of natural gas or raw biogas.

Alternatively, the methane in the offgas can be oxidised by a low-calorific combustor or by catalytic combustion. A number of manufacturers already provide applicable technologies on a commercial basis. These systems provide stable combustion even at methane concentrations as low as 3% in the combustion mixture with air. The treatment of offgas containing even less methane is increasingly difficult, as not enough energy is provided during the combustion of this gas, and raw biogas or biomethane have to be added in order to reach a stable oxidation. This is why it does not always make sense to choose an upgrading technology with a methane recovery as high as possible, because you always have to deal with the offgas. The integration of the upgrading plant into the biogas production facility and the overall concept of the biomethane production site are much more important. Only a very few upgrading technologies with extremely high methane recovery rates provide an offgas that is permitted to be directly vented to the atmosphere.

Production of Aviation Biofuel

Process and Technology Status

The aviation sector requires dropin bio-jet fuels that are functionally equivalent to fossil fuel and that are fully compatible with the existing infrastructure.

Jet fuel is a high-specification fuel that must meet standards as defined in ASTM D6155 for Jet A1, or by the Ministry of Defence Standard 91-91 in the UK. Bio-jet fuel must meet these same standards and in addition have an ASTM D7566 certification.

In addition to the four pathways that certified to date, several others are in the process, such as hydrotreated depolymerised cellulosic jet and green diesel (HDRD or HRD). The ASTM certification procedure is rigorous and can take years and millions of U.S. dollars to complete.

The four certified pathways to produce bio-jet fuels are described here and shown in figure below:

- HEFA: oleochemical conversion processes, such as hydroprocessing of lipid feedstocks obtained from oilseed crops, algae or tallow.

- FT: thermochemical conversion processes, such as the conversion of biomass to fluid intermediates (gas or liquid) followed by catalytic upgrading and hydro-processing tohydrocarbon fuels.

- SIP: biochemical conversion processes, such as the biological conversion of biomass (sugars, starches or lignocellulose-derived feedstocks) to longer chain alcohols and hydrocarbons.

- ATJ: A fourth category includes "hybrid" thermochemical or biochemical tech-nologies; the fermentation of synthesis gas; and catalytic reforming of sugars or carbohydrates.

For technologies of any type, a technology readiness- level scale is often used to deter-mine the maturity of a technology and its closeness to market. The Commercial Avia-tion Alternative Fuels Initiative's fuel readiness level (FRL) scale performs this func-tion for bio-jet pathways. The scale rates ASTMcertified fuels at FRL7 or higher.

A recent study by Mawhood evaluated pathways to bio-jet against this scale and con-cluded that HEFA was at FRL9, FT at FRL7 to FRL8, SIP between FRL5 and FRL7, and others in a range from four to six. Based on this scale, the recent ASTM certification of ATJ would give it a rating of FRL7. What these ratings do not capture, however, is the commercial viability of a certain pathway. As noted earlier, FT bio-jet was certified based on using coal as the feedstock, not biomass. While the feedstock does not affect fuel properties, it does have an impact on production potential, production.

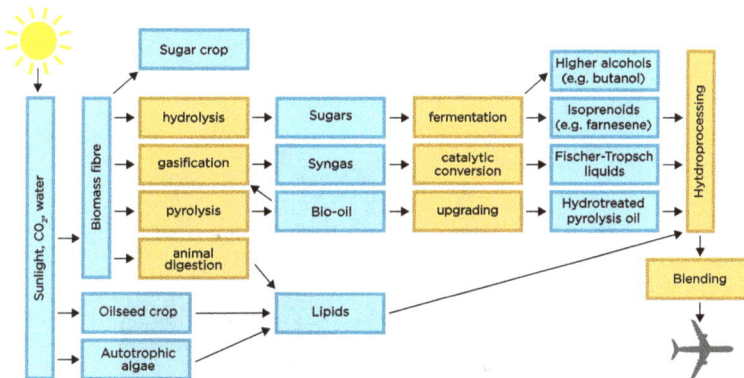

A simplified schematic diagram of different technology pathways to bio-jet fuel cost and technology readiness (RAND Corporation and Massachusetts Institute of Technology, 2009). It will also significantly impact the potential of the fuel to reduce GHG emissions. This leaves questions about the FT pathway, for which certification used coal as a feed-stock. This is the closest to commercialisation among the advanced pathways based on the FRL classification, but significant challenges still must be resolved if the method is to be useful in meeting emissions-reduction targets. A recent report by France's Académie des Technologies and Académie de l'Air et de l'Espace concluded that vegetable oil-based HEFA bio-jet is likely to be the only economically viable option in the near future.

A key aspect of bio-jet production is the requirement for hydrogen (H2) to upgrade oxygen-rich carbohydrate, lignin or lipid feedstocks to hydrogenrich hydrocarbons that are functionally equivalent to petroleum-derived jet fuel. Thus, some type of hydroprocessing step will likely be required for most bio-jet fuel technology platforms, with external sources of hydrogen used to remove oxygen in the form of water from the starting material, or to saturate double bonds in a final polishing step. Processes that do not require hydrogen can be used, including chemical and biological processes, but yields will be smaller because of the consumption of a portion of the feedstock.

The amount of hydrogen needed to produce bio-jet from a feedstock is illustrated by the effective hydrogen to carbon ratio, Heff/C, in the biomass feedstocks. This hydrogen-to-carbon ratio provides a useful metric to better understand and compare the technical and economic challenges of the various drop-in biofuel processes using different types of biomass feedstocks. This staircase approach, which involves assessing how much oxygen needs to be displaced by hydrogen, establishesa ranking of feedstocks by how "easy" they are to upgrade. This shows that oils and fats need the least hydrogen, and are therefore the easiest, whereas sugars and lignocellulosic biomass need the most.

The amount required is important to determine the overall impact on GHG emissions because most hydrogen is produced by the reforming of natural gas, making fossil-fuel consumption a part of the process.

Technologies for Oloeochemical Conversion Processes

The dominant HEFA pathway uses oil and fat feedstocks such as palm oil, used cooking oil and tallow. These "fatderived" or "oleochemical-derived" drop-in biofuels are often referred to as HEFA, but are also called hydrotreated vegetable oil biofuels (HVO). This technology is mature and currently operates at a commercial scale.

HEFA is notably distinct from fatty acid methyl ester (FAME) biodiesels, which retain an oxygen ester and are therefore too oxygenated to be used as a drop-in biofuel.

Almost all of the world's commercial plants that are able to make HEFA are currently producing mostly HEFA diesel; but as of 2016, California's AltAir Fuels had become the first dedicated HEFA biojet production facility. Notably, AltAir's production process results in a mixture of hydrocarbon molecules from which several products, such as renewable diesel, bio-jet and naphtha, must be separated.

The two main technologies used commercially to make HEFA are Neste's NEXBTL and UOP and Eni's EcofiningTM processes. The EcofiningTM technology has been licensed by several companies and is used in a number of facilities.

Vegetable oils contain about 10 wt% oxygen, which must be removed to produce drop-in HEFA biofuels. Hydrotreating of vegetable oils to remove this oxygen typically consumes about 3 wt% of hydrogen. Alternative deoxygenation processes that require less or no hydrogen also produce less HEFA because of carbon losses as part of the process.

Trials have shown that HVO biofuels can also be produced by co-processing oleochemical feeds with petroleum

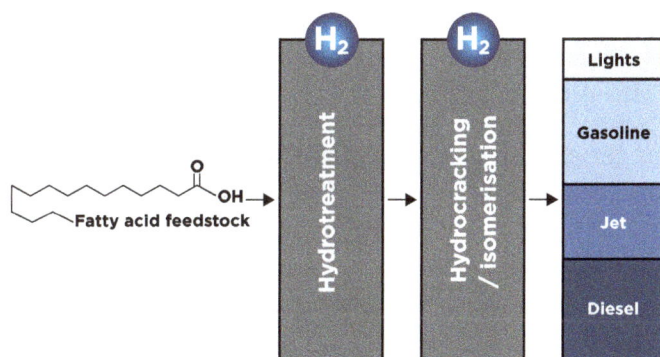

Basic Diagram of the Oleochemical Conversion Pathway feeds in modern oil refineries, although challenges remain. Chevron, Phillips66 and BP have an ASTM D7566 application underway to certify jet fuel produced through the co-processing of vegetable oils using this method at a 5% concentration.

The vast majority of the drop-in biofuels in aviation trials have used HEFA biojet primarily because the technology is mature and ASTM approved, but unless there are incentives to offset the high price, oil- and fat-based feedstocks are more likely to be converted to conventional FAME biodiesel. Another obstacle to scaling up the HEFA pathway is that increasing the fraction of jet-fuel range products produced from oleochemical feedstocks requires higher hydrogen inputs (more extensive hydrocracking) and also results in lower yields. These extra costs must be considered when developing strategies to promote increased bio-jet production. The proposed use of HEFA diesel as a bio-jet blend will minimise additional processing costs.

Vegetable-oil feedstocks require more land to produce than other feedstock types. Algae and non-food crops such as camelina, grown as a rotation crop, are examples of oleochemical feedstocks that require less land. These have been assessed by various initiatives, such as the U.S. Department of Agriculture "farm-to-fly" program; the U.S.

Department of Energy's National Alliance for Advanced Biofuels and Bioproducts consortium, which is now complete; the European 7th Programme for Research and Innovation; Horizon 2020; and the "Flightpath" programs. Cooking oil and tallow are much more sustainable, but availability is limited. In the short- to mid-term neither have the potential to contribute significantly to any increase in the production of bio-jet. The U.S. Environmental Protection Agency has estimated that about 3 bln gallons (11.4 bln L) of used cooking oil is produced each year in the U.S., resulting in a theoretical yield of about 9 bln L of HEFA bio-jet, based on 1.2:1 feedstock to fuel ratio. However, much of this feedstock is currently used to make biodiesel.

One strategy for improving costs of bio-jet based on the oleochemical conversion route is the use of HDRD diesel as a jet substitute. This also involves the upgrading of vegetable oils. Boeing and Neste have applied for ASTM certification under ASTM D7566 to blend small amounts of vegetable oil with ordinary jet fuel20. Boeing completed a test flight using a 15% blend of HDRD in December 2014. Although no information is available on costs, this blend should be cheaper to produce than HEFA bio-jet, as the blend is likely to require less processing and result in potentially higher yields. The process is similar to current HEFA diesel production.

It differs in that longer carbon-chain lengths from vegetable oils are not cracked into shorter jet-range molecules. Instead increased isomerisation is used to improve the cold flow properties of the blend, allowing low bio-jet blends that still meet the overall specifications for jet fuel.

Thermochemical Routes to Turn Biomass Into bio-jet fuel

The major challenges for thermochemical routes to bio-jet differ mainly in conversion-process efficiency and technology risks, although feedstock choices can also result in qualities in the end product. Thermochemical routes used to turn biomass into bio-jet involve the production of three main products, in different ratios: bio-oil, synthesis gas and char. The two main thermochemical routes to bio-jet are gasification and pyrolysis, and hydrothermal liquefaction (HTL). The FT process uses gasification combined with synthesis to produce bio-jet. Several commercial facilities based on gasification-FT are planned, and this pathway is discussed in more detail below. The pyrolysis route to biojet is known as HDCJ (hydrotreated depolymerised cellulosic jet). An ASTM application for HDCJ was initiated by KiOR, but the company is now in bankruptcy, creating a setback for the certification of this pathway.

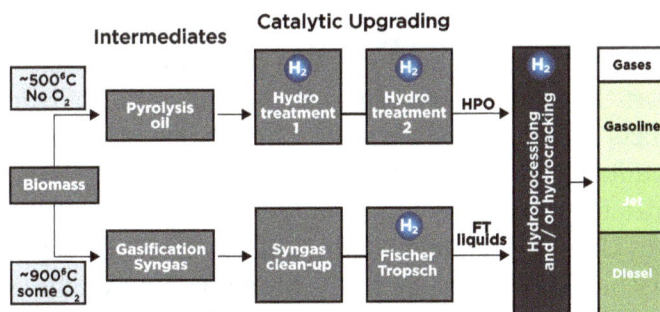

Thermochemical conversion routes

Gasification

Gasification involves the heating of small feedstock particles at high temperatures in a controlled-oxygen environment to produce synthesis gas, which is comprised of mostly H_2 and carbon monoxide and typically called syngas. Syngas converts to numerous gaseous and liquid chemicals or fuels via the FT process using catalysts. This process produces a mixture of hydrocarbon molecules from which various fuels and chemicals can be extracted.

Gasification and FT have been used since the 1980s by South Africa's Sasol company to convert coal into fuels, at a current capacity of 160 000 barrels per day (b/d). The FT process is also used in the world's largest natural gas-to-liquids plant, Shell's Pearl facility in Qatar. It was completed in 2011 and produces 140 000 b/d of fuels (IEA Bioenergy, 2014). Although bio-jet could potentially be produced via this process, using biomasses such as black liquor or bio-oil as a feedstock, various challenges have prevented this on a commercial basis.

Bio-jet produced via FT was certified by ASTM in 2009, at up to 50% of a blend with conventional jet fuel, but it was produced using coal as the feedstock and not biomass. The feedstock that is used to make FT syngas can result in different products during and after gasification. This significantly influences the composition of the syngas. For efficient fuel production, a desirable syngas should be a mixture of H2 and CO only, with all contaminants removed.

Gasification of biomass typically results in considerable tar formation that needs to be cleaned up, and the high oxygen content of biomass impacts the ratio of H_2 to CO in the synthesis gas. As a result, biomass-derived syngas is less energy dense than syngas derived from natural gas, has a lower ratio of hydrogen to carbon, and contains more impurities. Typically, biomass and MSW-derived syngas needs to be enriched in hydrogen and cleaned of impurities such as tars, nitrogen and other atoms comprised of anything other than carbon or hydrogen. These impurities can deactivate the synthesis catalysts. Although cleaning syngas is technically possible, it has proven to be costly. Plasma gasification is another way to produce a very clean syngas, but has also proven to be significantly more expensive.

To date, gasification technologies have entailed high capital costs to both gasify the biomass and convert the resulting syngas to FT liquids or partially oxygenated liquid hydrocarbon products such as mixed alcohols. These types of coalor natural gas-fed facilities have been built at a large scale in the hopes of capturing economies of scale. The capital cost estimates for a first-of-itskind commercial gasification-based facility range from USD 600 m to USD 900 m, and would typically have the capacity to produce 2 000 t per day of dry biomass. Although this is a significantly smaller size than current facilities based on coal and natural gas, logistics challenges will be likely because biomass is a less energy-dense feedstock.

Some of these anticipated supply-chain challenges can be mitigated through the use of alternative feedstocks such as pyrolysis bio-oils, which are more energy dense than wood. Because a range of hydrocarbon molecules are produced by the FT process, large-scale facilities are also in a better position to market the multiple commercial products created, thus improving the overall economics.

Current FT technology results in a maximum of about 40% of the final product comprised of bio-jet fuel and middle distillates, requiring the marketing of the other 60% of the output.

Commercial biomass-gasification facilities under construction include those of Fulcrum Bioenergy and Red Rock Biofuels, in U.S. Kaidi has proposed building an FT facility in Finland. These pioneer plants should provide invaluable insights and lessons for future investment. Many people believe that costs for the FT route could fall considerably as the technology matures. Fulcrum Bioenergy claims to be able to produce FT transportation fuel at less than USD 1 per gallon (USD 0.26/L) using MSW as a feedstock.

A proposal from Solena and British Airways to use MSW in a gasification process was recently cancelled. Solena hoped to use plasma gasification technology that was likely to cost significantly more than what Fulcrum will use. Red Rock Biofuels plans to use woody biomass as well as a different FT technology (from Velocys).

Pyrolysis and Hydrothermal Liquefaction (HTL)

Fast pyrolysis exposes small biomass particles of about 3 mm in length to heat at 500°C for a few seconds to produce a bio-oil with up to 75 wt% yield. Although companies such as Ensyn in Canada have been producing fast-pyrolysis biooils for many years, these have mainly been used in niche applications such as food flavouring.

Energy applications have been restricted to heavy fuel oil used in stationary heating, and in powergenerating facilities. Although Ensyn22 recently obtained regulatory approval for RFDiesel and RFGasoline, which are fuel products generated via co-processing in petrochemical refineries, no jet fuel has been produced this way.

In the Netherlands, BTG has commercialised the flash pyrolysis technology in its EMPYRO23 project. As of late 2016, however, biooil was used to replace natural gas in a heating application in a milk factory. BTG has also been testing possible co-processing of bio-oil in a petroleum refinery.

The commercial production of bio-jet via the pyrolysis route is likely to be challenging because biocrudes derived from fast pyrolysis contain up to 40% oxygen, similar to the biomass itself. This necessitates extensive upgrading to produce bio-jet, which is typically achieved through hydroprocessing. These processing costs, as well as the need for external hydrogen, represent a large proportion of equipment and production costs.

A further challenge to the hydroprocessing of pyrolysis oils is the cost and stability of the catalysts that are required.

A potential advantage of the pyrolysis approach to bio-jet production is that it can be done in existing oil refineries, which reduces the need for capital to build a dedicated facility.

Similarly, significant savings might be achieved by directly sourcing hydrogen from an oil refinery and, in the longer term, through using existing processing units.

Co-processing in existing petroleum refineries is considered a key strategy for upgrading pyrolysis-derived bio-oils, but comes with some technical challenges. These include selecting the point of insertion, the extent to which upgrading is required prior to insertion and the disparate types of catalysts needed for bio-oils compared with those used in oil refining. Refinery-insertion strategies should be synergistically beneficial but are likely more technically challenging than is generally acknowledged.

Catalytic pyrolysis or processes such as HTL can produce a bio-oil intermediate with significantly lower oxygen content, at less than 10%. That would be easier to upgrade to produce fuels, including bio-jet.

Although some studies have indicated that this method could potentially produce the lowest-cost bio-jet, the high-pressure requirements of HTL during the production of biocrude will impact their potential for scale-up. While production of bio-oil via pyrolysis is at a commercial scale, HTL is currently just at the demonstration stage, as pioneered by Licella's Australian plant. Although there is a scarcity of reliable technical and economic analyses, a minimum fuel-selling price (MFSP) of USD 3.39 per gallon could be achieved when making diesel and gasoline via fast pyrolysis followed by upgrading.

Biochemical Routes to Turn Biomass Into Bio-jet Fuel

Capital expenditure for biochemical routes to bio-jet are projected to be lower than those for thermochemical routes, but that benefit may be offset if the MFSP is higher, which is also expected (table). In contrast to the more familiar sugar-to-ethanol fermentation route to bioethanol, advanced biological routes convert sugars to less oxygenated and more energydense molecules such as longer-chain alcohols like butanol and butanediol. Advanced biological routes can also convert sugars to larger hydrocarbon molecules such as isoprenoids and fatty acids. Amyris, using genetically engineered yeasts, converts sugars directly to renewable hydrocarbons such as farnesene. Farnesene can then be upgraded to farnesane through hydroprocessing, producing SIP. It is also known as the Direct Sugars to Hydrocarbons (DSHC) pathway. It received ASTM certification in 2014, provided it is used in a 10% blend with fossil-derived jet fuel.

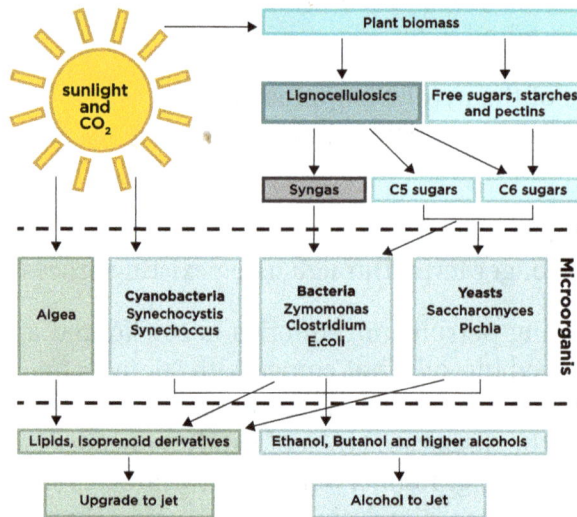

From microorganism to bio-jet fuel with various renewable feedstocks

Although butanol, n-butanol and isobutanol are oxygenated and thus cannot be considered to be fully "dropin" biofuels, they are less oxygenated and less hydrophilic than ethanol and can be used to produce bio-jet fuel through the ATJ pathway. Gevo recently obtained ASTM certification for its bio-jet fuel produced from isobutanol, which Alaska Air has used in a commercial flight.

The potential to use existing ethanol facilities lies in the ability to replace the existing yeasts that produce ethanol with alternative microorganisms that could instead make bio-jet. Because some of the biological intermediates such as farnesene are quite hydrophobic they should in theory be more readily recoverable from the aqueous fermentation broths. Nevertheless, the recovery of these molecules from the fermentation broth has been more challenging than predicted because of intracellular expression of hydrophobic metabolites. For butanol, fermentation titers are typically well below the concentration that induces phase separation, which ranges from 70 g/L of butanol to 80 g/L.

Biochemical-based bio-jet pathways can produce highly reduced biohydrocarbon molecules such as sesquiterpenes and fatty acids based on the used microorganism, which accordingly decrease the degree of final hydroprocessing required to meet jet-fuel specifications. Even still, advanced biological pathways requiremore energy and carbon-intensive metabolic processes than required in ethanol production in order to decrease feedstock sugars.

Given the potential commercialisation scale, conventional ethanalogenic yeast fermentations could achieve a higher order of magnitude than the product yield for fermentation-derived biojet. Furthermore, biochemical-based bio-jet platforms are relatively pure, and functionalised long carbon-chain molecules can be produced. This is an advantage over thermochemical-based processes.

The production of bio-jet would not be the most profitable use of biochemically processed biomass and sugars. Products such as carboxylic acids, alcohols and polyols can generate higher profits because there are fewer processing steps, lower NADPH requirements, and less hydrogen consumption. Less-oxygenated microbial metabolites with potential as drop-in biofuel intermediates are already being sold in the value-added chemicals and cosmetics markets, such as Amyris's farnesene and Gevo's or Butamax's butanol. The market for biochemical drop-in products is highly competitive and growing. The volume of biochemical drop-in products in the market is expected to increase by between 10 Mt and 50 Mt annually to 2020, which is equal to the current markets size of biofuels. Incentives will be required for commercial entities to concentrate on drop-in fuels until the market is saturated.

Hybrid Conversion Processes

Hybrid conversion processes can combine a mixture of the above methods. For example, the ATJ process can combine biochemical production of alcohol, through fermentation of sugars or conversion of syngas from gasification and catalytic conversion of the alcohol to bio-jet. Another example of a hybrid process is Virent's aqueous phase reforming. This process uses feedstock sugars potentially derived through biochemical conversion of lignocellulosic feedstocks, while synthesis takes placevia catalysis.

Methods used in Production

Supercritical Process

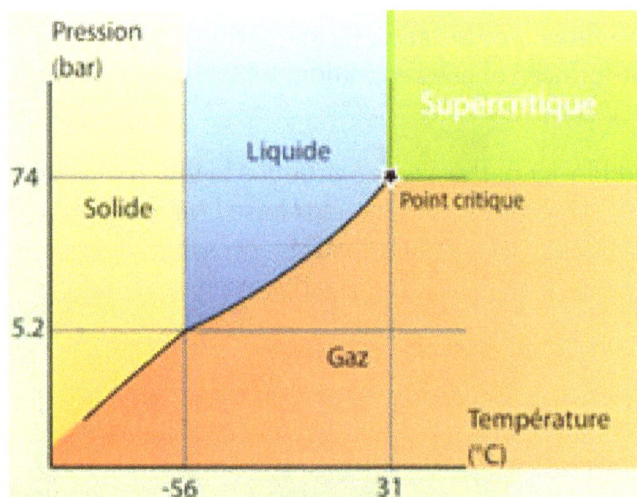

Any substance is characterized by a critical point which is obtained at specific conditions of pressure and temperature. When a compound is subjected to a pressure and a temperature higher than its critical point, the fluid is said to be "supercritical".

In the supercritical region, the fluid exhibits particular proporties and has an intermediate behavior between that of a liquid and a gas. In particular, supercrical fluids (SCFs) possess liquid-like densities, gas-like viscosities and diffusities intermediate to that of a liquid and a gas.

As shown on the phase diagram, the gas-liquid equilibrium curve is interrupted at the critical point, providing a continuum of physico chemical properties.

The fluid is said "supercritical" when it is heated obove its crititcal temperature and compressed obove its cirical pressure. This particular behavior of substances was first observed in 1822 by French engineer and physicist, Charles Cagniard de La Tour in his famous cannon barrel experiment. It was then defined as supercritical fluid by Irish chemist, Thomas Andrews. The most widely used supercritcal fluids are CO_2 and water.

Supercritical fluids based processes include extraction, impregnation, particle formation, formulation, sterilization, cleaning and chemical reactions among others. In all cases, the supercritical fluid is used as an alternative to traditional organic liquid solvents. The most widely used supercritical fluids are CO_2 (Tc – 31°C, Pc = 74 bar) and water (Tc – 374°C, Pc = 221 bar) but some proccsscs (extraction, reactions) involve the use of supercritical methanol, ethanol, propane, ethane mainly. Supercritical CO_2 processes are the most widely spread as these are exempt of the operations of elimination of solvent residues, operations generally needed when the solvent used is an organic compound.

Some examples of proceseses carried out under supercritical and subcritical conditions are listed below.

Extraction

The principle of the extraction of solids using supercritical CO_2 relies on the strong variation of the solvation power of CO_2 which occurs by simple variation the operating conditions (temperature and pressure). This allows to selectively extract molecules according to their chemical nature. Very weakly polar, CO_2 turns out to be an excellent solvent of nonpolar or small polar molecules in supercritical conditions. Once the desired compound is dissolved in supercritical CO_2, the pure extract can easily be obtained by simple depressurization. This results in the separation of CO_2, which becomes gas again, from the extract, which is recovered in solid or liquid form. In theory, any porous solid material (plants, plastics, wood, can be treated by supercritical fluid extraction to recover valuable compounds (oils, fragrances, pigments) or undesired substances (pollutants, residual solvents).

Subcritical water extraction can also be carried out to extract hydrophobic compounds (polyphenols, terpenes, tannins) from various plant materials (fruits, wood, algae).

Fractionation

Based on the principle of supercritical CO_2 extraction of solids, the extraction (fractionation) of liquids can also be performed in order achieve high levels of purification. Examples of liquids concerned by this process include plant extracts, vegetable oils, fish oil, polymers.

Impregnation

The principle of supercritical fluid impregnation consists of a scan of a porous solid material (polymers, wood, textiles) by a supercritical (mainly CO_2) in which the active substance to impregnate is previously dissolved. This step is followed by a relaxation that causes the passage of the CO_2 in the gaseous state, leaving the "target" material impregnated with the active substance. Therefore, the impregnation of a solid matrix with an active compound is easy achievable using supercritical fluids based technologies. Example of the use of supercritcal impregnation include the dyeing of textile, the tanning of leather, wood impregnation.

Powder Formation

This refers to micro and nanoparticle generation, crystalization, precipitation, micronization of inorganic, organic, pharmaceutical, and polymeric materials.There are two main ways of precipitating micro and nanoparticles either using supercritical fluid as solvent, the RESS technique (Rapid Expansion of Supercritical Solutions); or using it as anti-solvent, the SAS technique (Supercritical Anti-Solvent). Encapsulation of variuous substances and polymers can also be performed using the PGSS (Particles from Gas Saturated Solutions) process.

Deposition

Chemical Fluid Deposition (CFD) involves the chemical reduction of organometallic compounds in supercritical fluids to yield high purity deposits. Typically, the reaction is initiated upon the addition of H_2 or other reducing agent. The advantages of CFD over conventional deposition techniques are a consequence of the unique properties of supercritical fluids, thus promoting infiltration into complex geometries and minimizing mass transfer limitations common to liquid phase reductions. High purity metal films including Pt, Pd, Au, Rh, Ni, Cu, Al have been deposited by CFD from supercritical fluids CO_2 using appropriate precursors.

Chemical Reactions

Conducting chemical reactions at supercritical conditions affords opportunities to manipulate the reaction environment (solvent properties) by increasing pressure to enhance the solubilities of reactants and products, to eliminate interphase transport limitations thus increasing reaction rates, and to integrate reaction and separation unit operations. Supercritical conditions may be advantageous for reactions involued in fuels processing, biomass conversion, biocatalysis, homogeneous and heterogeneous catalysis, environmental control, polymerization, materials synthesis and chemical synthesis. Examples of chemical reactions carried out at an industrial scale include hydrogenation, oxydation, esterification and etherfication reactions among others.

Sterilization

Supercritical fluids have demonstrated the ability to inactivate bateria, fungi, yeasts and viruses thus providing an efficient mean for the sterilization of food and medical devices. The mechanism of micro-organism inactivation are not fully understood yet but some evidence show that this may be caused by cell wall alteration due to a strong interaction of the fluid with the lipids and the inactivation of key-enzymes resulting from pH decrease inside the cell.

Drying

Supercritical drying processes rely on the extraction of water and other solvents using supercritical CO_2. The absence of surface tension allows the supercritical fluid to be removed without distortion. Such processes are used to make aerogels but also for the lyophilization (freeze-drying) of biological and food matrices and the dry-cleaning of clothes (as a replacement for chlorinated solvenst) in an environmentally-friendly way.

Cleaning

In the cleaning process, the use of supercritical CO_2 with or without the addition of specific surfactants avoids the use of solvents such as trichloroethylene. For environmental reasons, this toxic solvent is subject to many health and regulatory limitations, leaving the place to "green" solvents. Although only hydrocarbon solvents are nowadays able to meet the market's demand for alternative solutions to chlorinated solvents, the use of CO_2 under pressure or supercritical CO_2 appears as a particularly interesting industrial alternative to chlorinated solvents. This process is industrialized for the dry-cleaning of textiles, the cleaning of mechanical spare parts.

Chromatography

Supercritical Fluid Chromatography (SFC) is used for the analysis and purification of low to moderate molecular weight, thermally labile molecules and the the separation of chiral compounds. Principles are similar to those of high performance liquid chromatography (HPLC), however SFC typically utilizes carbon dioxide as the mobile phase. Therefore the technique is more versatile, exhibits better resolution and faster analysis times than conventional liquid chromatographic methods.

Refrigeration

Supercritical CO_2 can be used as a refrigerant in air-conditioning and cooling system providing eco-friendly solutions in industrial and domestic heat-pumps.

Ultra- and high-shear in-line and Batch Reactors

Ultra- and High Shear in-line or batch reactors allow production of biodiesel continuously, semicontinuously, and in batch-mode. This method drastically reduces production time and increases production volume. The reaction takes place in the high-energetic shear zone of the ultra- and high Shear mixer by reducing the droplet size of the immiscible liquids such as oil or fats and methanol. Therefore, the smaller the droplet size, the larger the surface area the faster the catalyst can react.

Ultrasonic Reactor Method

In the ultrasonic reactor method, the ultrasonic waves cause the reaction mixture to produce and collapse bubbles constantly. This cavitation provides simultaneously the mixing and heating required to carry out the transesterification process. The ultrasonic reactor method for biodiesel production drastically reduces the reaction time, reaction temperatures, and energy input. Industrial scale ultrasonic devices allow for the industrial scale processing of several thousand barrels per day.

Ultrasonic cavitation provides the necessary activation energy for the industrial biodiesel transesterification.

Transesterification

Manufacturing biodiesel from vegetable oils (e.g. soy, canola, jatropha, sunflower seed or algae) or animal fats, involves the base-catalyzed transesterification of fatty acids with methanol or ethanol to give the corresponding methyl esters or ethyl esters. Glycerin is an inevitable byproduct of this reaction.

Vegetable oils as animal fats are triglycerides composed of three chains of fatty acids bound by a glycerin molecule. Triglycerides are esters. Esters are acids, like fatty acids, combined with an alcohol. Glycerine (= glycerol) is a heavy alcohol. In the conversion process triglyceride esters are turned into alkyl esters (= biodiesel) using a catalyst (lye) and an alcohol reagent, e.g. methanol, which yields methyl esters biodiesel. The methanol replaces the glycerin.

The glycerine the heavier phase will sink to the bottom. Biodiesel the lighter phase floats on top and can be separated, e.g. by decanters or centrifuges. This conversion process is called transesterification.

The conventional esterification reaction in batch processing tends to be slow, and phase separation of the glycerin is time-consuming, often taking 5 hours or more.

Ultrasonics for Biodiesel Processing

Today, biodiesel is primarily produced in batch reactors. Ultrasonic biodiesel processing allows for the continuous inline processing. Ultrasonication can achieve a biodiesel yield in excess of 99%. Ultrasonic reactors reduce the processing time from the conventional 1 to 4 hour batch processing to less than 30 seconds. More important, ultrasonication reduces the separation time from 5 to 10 hours (using conventional agitation) to less than 60 minutes. The ultrasonication does also help to decrease to amount of catalyst required by up to 50% due to the increased chemical activity in the presence of cavitation. When using ultrasonication the amount of excess methanol required is reduced, too. Another benefit is the resulting increase in the purity of the glycerin.

Ultrasonic processing of biodiesel involves the following steps:

- The vegetable oil or animal fat is being mixed with the methanol (which makes methyl esters) or ethanol (for ethyl esters) and sodium or potassium methoxide or hydroxide

- The mix is heated, e.g. to temperatures between 45 and 65degC

- The heated mix is being sonicated inline for 5 to 15 seconds

- Glycerin drops out or is separated using centrifuges

- The converted biodiesel is washed with water.

Most commonly, the sonication is performed at an elevated pressure (1 to 3bar, gauge pressure) using a feed pump and an adjustable back-pressure valve next to the flow cell.

References

- Chemistry-of, biodiesel: goshen.edu, Retrieved 29 May 2018

- Biofuels-status-and-perspective/an-overview-of-bioethanol-production-from-algae: intechopen. com, Retrieved 19 May 2018

- Irena-biofuels-for-aviation-2017: irena.org, Retrieved 31 March 2018

- Supercritical-fluids.146.0: supercriticalfluid.org, Retrieved 21 May 2018

- Biodiesel-transesterification-01: hielscher.com, Retrieved 18 July 2018

Permissions

All chapters in this book are published with permission under the Creative Commons Attribution Share Alike License or equivalent. Every chapter published in this book has been scrutinized by our experts. Their significance has been extensively debated. The topics covered herein carry significant information for a comprehensive understanding. They may even be implemented as practical applications or may be referred to as a beginning point for further studies.

We would like to thank the editorial team for lending their expertise to make the book truly unique. They have played a crucial role in the development of this book. Without their invaluable contributions this book wouldn't have been possible. They have made vital efforts to compile up to date information on the varied aspects of this subject to make this book a valuable addition to the collection of many professionals and students.

This book was conceptualized with the vision of imparting up-to-date and integrated information in this field. To ensure the same, a matchless editorial board was set up. Every individual on the board went through rigorous rounds of assessment to prove their worth. After which they invested a large part of their time researching and compiling the most relevant data for our readers.

The editorial board has been involved in producing this book since its inception. They have spent rigorous hours researching and exploring the diverse topics which have resulted in the successful publishing of this book. They have passed on their knowledge of decades through this book. To expedite this challenging task, the publisher supported the team at every step. A small team of assistant editors was also appointed to further simplify the editing procedure and attain best results for the readers.

Apart from the editorial board, the designing team has also invested a significant amount of their time in understanding the subject and creating the most relevant covers. They scrutinized every image to scout for the most suitable representation of the subject and create an appropriate cover for the book.

The publishing team has been an ardent support to the editorial, designing and production team. Their endless efforts to recruit the best for this project, has resulted in the accomplishment of this book. They are a veteran in the field of academics and their pool of knowledge is as vast as their experience in printing. Their expertise and guidance has proved useful at every step. Their uncompromising quality standards have made this book an exceptional effort. Their encouragement from time to time has been an inspiration for everyone.

The publisher and the editorial board hope that this book will prove to be a valuable piece of knowledge for students, practitioners and scholars across the globe.

Index

H
Hydrolysis, 41, 63, 101, 107-109, 112, 118, 124-125, 141-142, 144-153

I
Impregnation, 170-171
Internal Combustion Engine, 4, 7, 15, 23, 53

J
Jatropha, 11, 26-27, 31, 87, 97, 99, 116-117, 130, 133, 174
Jet Fuel, 12, 29-30, 84, 160-168

L
Landfill, 3, 8, 10-11, 38-39, 55, 60, 159
Lignin, 35, 64-65, 67, 99-101, 107, 109-110, 141, 144-145, 148-151, 162

M
Marginal Land, 10-11, 71
Membrane Technology, 158
Methane, 8, 12, 17-18, 37-41, 45, 50, 54-55, 64, 66, 111, 153-160
Methanol, 8, 16-19, 24-25, 30-33, 39, 42, 54, 94, 96, 104-105, 115, 118-119, 128-129, 131, 133, 136-139, 141, 150, 170, 173-174
Municipal Solid Waste, 3, 11-12, 100

O
Offgas, 154-157, 159-160

P
Paper Waste, 19, 123-124
Poplar, 57, 60, 66-72, 86, 100, 148
Propanol, 8, 24-25, 105

P (col2)
Purification, 33, 40-41, 105, 118, 120, 123, 136, 139-141, 153, 171, 173
Pyrolysis, 8-9, 35, 52, 54, 63, 65, 80, 113, 147, 161, 165-168

R
Rapeseed, 11, 31, 87, 97, 114, 117-120, 126, 133-134, 139
Renewable Energy, 5, 16, 21, 38, 51, 58, 66, 73, 81, 90, 92, 96

S
Saccharification, 80, 107-110, 112, 152
Soybean, 2, 4, 11, 31, 57, 72, 84, 89, 93-97, 100, 114-116, 121-122, 133-135
Sugarcane Bagasse, 100, 148
Sulphuric Acid, 124-125, 141-142, 149-150
Supercritical Fluid, 170-173
Syngas, 8-9, 15, 17, 35, 39-42, 63, 79, 101, 103, 113, 161, 165-166, 168-169

T
Thermochemical Conversion, 9, 80, 103, 161, 165
Transesterification, 4, 8, 30-33, 63, 94, 103, 105-106, 115, 118-119, 128-129, 136-141, 173-175

V
Vegetable Oil, 8-9, 11-12, 15, 31, 79, 94, 96-98, 120, 122-123, 127-129, 136, 162, 164, 174

W
Water Scrubbing, 154-155
Woody Biomass, 51-52, 57, 59-63, 66-67, 73-74, 77, 113, 166

X
Xylose, 99-100, 149

* 9 7 8 1 6 4 1 1 6 2 2 4 1 *